Lecture Notes in Mathematics

Edited by A. Dold and B. Eckmann

456

Robert M. Fossum
Phillip A. Griffith
Idun Reiten

Trivial Extensions of Abelian Categories

Homological Algebra of Trivial Extensions of Abelian Categories with Applications to Ring Theory

Springer-Verlag
Berlin · Heidelberg · New York 1975

Authors
Dr. Robert M. Fossum
Dr. Phillip A. Griffith
Department of Mathematics
University of Illinois
Urbana, Illinois 61801
USA

Dr. Idun Reiten
Matematisk Institutt
Universitetet I Trondheim
Norges Laererhøgskole
N-7000-Trondheim

Library of Congress Cataloging in Publication Data

Fossum, Robert M
Trivial extensions of Abelian categories.

(Lecture notes in mathematics ; 456)
Bibliography: p.
Includes index.
1. Commutative rings. 2. Associative rings.
3. Abelian categories. I. Griffith, Phillip A., joint author. II. Reiten, Idun, 1942- joint author. III. Title. IV. Series: Lecture notes in mathematics (Berlin) ; 456.
QA3.L28 no. 456 [QA2151.3] 510'.8s [512'.55]
75-12984

AMS Subject Classifications (1970): 13A20, 13C15, 13D05, 13H10, 16A48, 16A49, 16A50, 16A52, 16A56, 18A05, 18A25, 18E10, 18GXX

ISBN 3-540-07159-8 Springer-Verlag Berlin · Heidelberg · New York
ISBN 0-387-07159-8 Springer-Verlag New York · Heidelberg · Berlin

Offsetdruck: Julius Beltz, Hemsbach/Bergstr.

Contents

Introduction v

Section 0: Preliminaries 1

Section 1: Generalities 3

Section 2: Coherence 24

Section 3: Duality and the Gorenstein property 35

Section 4: Homological dimension in $\underline{\underline{A}} \ltimes F$ 52

Section 5: Gorenstein modules 87

Section 6: Dominant dimension of finite algebras 104

Section 7: Representation dimension of finite algebras 113

References 117

Introduction

The notion of the trivial or split extension of a ring by a bimodule has played an important role in various parts of algebra. In most cases, however, it is introduced *ab initio* and then used with a particular purpose in mind. With no intention of being comprehensive, we mention some important applications of this construction.

But first we must describe the construction. Suppose R is a ring (with identity) and M is an R-bimodule. The set $R \times M$, with componentwise addition and multiplication given, elementwise, by $(r,m)(r',m') = (rr',mr' + rm')$, becomes a ring, which we denote by $R \ltimes M$. It has an ideal $(0 \times M)$ which has square zero. And there is a ring homomorphism $R \longrightarrow R \ltimes M$ and an augmentation $\pi: R \ltimes M \longrightarrow R$.

Hochshild, in studying the cohomology of R with coefficients in M, notices that $R \ltimes M$ is the extension of R by M corresponding to the zero element in the 2nd cohomology group $H^2(R,M)$. And this is related to the fact that any derivation $d:R \longrightarrow M$ defines an automorphism of the ring $R \ltimes M$ which induces the identity when composed with the augmention. Conversely, any such automorphism defines a derivation. In fact the automorphism corresponding to d is the map given by $(r,m) \longmapsto (r,d(r) + m)$. On the other hand, if $a:R \ltimes M \longrightarrow R \ltimes M$ satisfies the property $\pi a(r,m) = r$, then the map $d:R \longrightarrow M$ given by $d(r) = a(r,o) - (r,o)$ is a derivation.

This relation between these special ring automorphisms and derivations is useful, for example, in algebraic geometry, where the tangent space to a K-scheme S is defined as the K-scheme morphisms

$$\mathrm{Hom}_{Sch/K}(\mathrm{Spec}(K \ltimes K),\ S).$$

In this particular situation the ring $K \ltimes K \cong K[x]/(x^2) \cong K[\delta]$ is called the ring of "dual numbers".

Nagata makes particularly good use of the construction. He calls it the "principle of idealization". Thus any module over a commutative ring can be thought of as an ideal in a commutative ring. So any result concerning ideals has an interpretation for modules. This is useful in the primary decomposition theory for commutative noetherian rings and finitely generated modules. In the section 5 on Gorenstein modules, this principle will be exploited extensively to show how the theory of Gorenstein modules, in particular the theory of canonical modules, can be reduced to the theory of Gorenstein rings; consequently, these results are almost immediate consequences of Bass' original theory of Gorenstein rings.

A particularly striking case where these rings arise is in the category of rings. Suppose R is a ring. Let $\underline{\underline{\mathrm{Ann}}}_R$ denote the category of rings over R. That is $\underline{\underline{\mathrm{Ann}}}_R$ has as objects ring homomorphisms $S \longrightarrow R$ with the obvious morphisms. Then the monoid objects coincide with the group and abelian group objects in this category, and these are just the trivial extensions of R by R-bimodules. This is a result due to Quillen [59]. Quillen discusses cohomology theory. Thus we see a return to the first mentioned application of this construction.

In this paper other examples of general constructions related to rings are seen to be of this form (e.g., triangular matrix rings and categories of complexes over rings).

To the best of our knowledge, no general treatment has been given attempting to relate the homological properties of the ring $R \ltimes M$ with those of the ring R and the bimodule M (before the appearance of our expository paper on the subject). Perhaps, because in the commutative (noetherian) case the ring $R \ltimes M$ always has infinite global dimension, (provided $M \neq (o)$) there seems to be no connection. But in case M is not a symmetric R-module, the ring $R \ltimes M$ can have finite global dimension. And even when M is symmetric as an R-module, and $R \ltimes M$ is a noetherian ring with finite Krull dimension, the finitistic projective dimensions are finite (Raynaud and Gruson [61]). So there are some interesting cases in which the relations can be studied.

The main purpose of this paper, then, is to study the relations, if any, between various homological properties of the objects R, M and $R \ltimes M$. We have in mind global dimension, finitistic projective dimension, change of rings theorems, Gorenstein properties and dominant dimension.

At this point we pause to mention the problems which we began to study and which led to the more general theory. Suppose Λ is a ring with finite global dimension, say gl.dim $\Lambda = n$. Then Eilenberg, Rosenberg and Zelinsky [17] showed that the ring of lower triangular $m \times m$ matrices, $T_m(\Lambda)$, has finite global dimension, gl.dim $T_m(\Lambda) = n + 1$. M. Auslander asked whether the finistic projective dimension, FPD, was also preserved in this fashion. And indeed we have shown that $\mathrm{FPD}(T_m(\Lambda)) = 1 + \mathrm{FPD}(\Lambda)$. In this same connection, Auslander asked whether $T_m(\Lambda)$ is k-Gorenstein if Λ is k-Gorenstein. We have answered this by showing that $T_m(\Lambda)$ is k-Gorenstein if and only if Λ is k-Gorenstein. Together with the results about global dimension, it is seen that there is no relation between Gorenstein properties and global dimension, contrary to the commutative case where a ring is (locally) Gorenstein if it has finite global dimension.

We found, in working with these problems, that an effective means for constructing projective and injective resolutions of modules over the triangular matrix rings, or over the trivial extension rings, was not available. But in our investigations we found a very general method for constructing these resolutions. Basically, the problem reduces to considering a module X over the ring $R \ltimes M$ as an R-homomorphism $f: M \otimes_R X \longrightarrow X$ (considering X as an R-module) satisfying the relation $f.M \otimes_R f = 0$. It is seen that this immediately generalizes to an abelian category $\underline{\underline{A}}$ equipped with an endofunctor $F : \underline{\underline{A}} \longrightarrow \underline{\underline{A}}$. We then study the morphisms $f : FX \longrightarrow X$ in $\underline{\underline{A}}$ such that $f. Ff = 0$.

We are now prepared to discuss in more detail the contents of our paper section by section.

The paper begins with a very short section which introduces our notations. We have adhered to standard notations in ring theory.

In section 1 we introduce our notion of a trivial extension $\underline{\underline{A}} \ltimes F$ of an abelian category $\underline{\underline{A}}$ by an endofunctor F. After defining the category and pairs of adjoint functors relating the category to $\underline{\underline{A}}$, we go on to discuss projective and injective objects in $\underline{\underline{A}} \ltimes F$ and give a complete determination of them in terms of data in $\underline{\underline{A}}$. We discuss minimal epimorphisms and essential monomorphisms. An immediately apparent feature is the "duality in statements" between right exact functors $F:\underline{\underline{A}} \longrightarrow \underline{\underline{A}}$ and left exact functors $G:\underline{\underline{A}} \longrightarrow \underline{\underline{A}}$. Thus, we can find the projective objects in $\underline{\underline{A}} \ltimes F$ and the injective objects in $G \rtimes \underline{\underline{A}}$. But when F is a left adjoint to G, then the categories $G \rtimes \underline{\underline{A}}$ and $\underline{\underline{A}} \ltimes F$ are isomorphic. The section is concluded by relating the general construction to the more specific trivial extension of a ring by a module and interpreting the results for these specific constructions.

Section 2 is devoted to studying the coherence of the trivial extension $\underline{\underline{A}} \ltimes F$ with respect to a family of projective objects and the relation to the coherence of $\underline{\underline{A}}$ and properties of the derived functors of F. As an application we get the following result: The ring $R \ltimes M$ is left coherent if and only if R is left coherent and, for every finitely presented left R-module A, the left R-modules $\mathrm{Tor}_i^R(M,A)$ are coherent for $i > o$ and, if B is a finitely generated left submodule of $M \otimes_R A$, then B is finitely presented and $M \otimes_R B$ is finitely generated. This generalizes a result due to Roos [65].

In section 3 we discuss Auslander's notion of a pseudoduality, the notion of a k-Gorenstein category (ring) and the Gorenstein property of the ring of lower triangular matrices. As applied to rings, the left and right coherent ring R is k-Gorenstein if, for all i in the range $1 \leq i \leq k$, for all finitely presented left R-modules A, for all finitely presented (or generated) right submodules B of $\mathrm{Ext}_R^i(A,R)$, and for all $j < i$, we have $\mathrm{Ext}_R^j(B,R) = 0$. Auslander has shown that this is a left-right symmetric condition. His proof is included here for completeness since it has not been published elsewhere. After the proof of Auslander's theorem, we show that R is k-Gorenstein if and only if $T_2(R)$ is k-Gorenstein. It is clear that the proof easily generalizes to general $m \times m$ lower triangular matrices. We also include an example which shows that the Gorenstein property is very unstable.

In section 4, we discuss the homological dimension of objects in $\underline{\underline{A}} \ltimes F$ (and $G \rtimes \underline{\underline{A}}$). For a more complete description of the results in this section, you are referred to the introduction of the section, since the details are most precisely stated for special objects in $\underline{\underline{A}} \ltimes F$. In our expository paper [22], we were able to give results concerning the global dimension and the finitistic projective dimension of categories of the form $(\underline{\underline{A}} \times \underline{\underline{B}}) \ltimes \tilde{F}$, where $F : \underline{\underline{A}} \longrightarrow \underline{\underline{B}}$ is a right exact functor and $\tilde{F}(A,B) = (0,FA)$. The prototype of such a category is the category of (left) modules over a triangular matrix ring $\begin{pmatrix} R & 0 \\ {}_SM_R & S \end{pmatrix}$. Then Palmer and Roos [56,57] made a nearly complete determination of the situation in which left gl.dim $(R \ltimes M) < \infty$. Their results, in part, are stated in terms of spectral sequences. One of our aims in section 4 (Part A) is to use our own techniques (making only the mildest use of spectral sequences) to provide rather simple criteria under which $\mathrm{FPD}(\underline{\underline{A}} \ltimes F)$ and gl.dim $(\underline{\underline{A}} \ltimes F)$ are finite. In one of several examples which illustrate our technique, we give a simple construction of a triangular matrix ring Λ such that (left gl.dim Λ)-(right gl.dim Λ) $= m + 1$, where m is an arbitrary positive integer (such examples have been constructed by Jategaonkar [35]).

Near the end of section 4 (Part D), we consider conditions under which $R \ltimes M$ has finite (left) self injective dimension (where R is a ring and M is an R-bimodule). These results are applied in section 5 (Gorenstein modules).

In Section 5, we study those (local) Noetherian commutative rings A and finitely generated A-modules M for which the ring $A \ltimes M$ is Gorenstein. Our starting point is a result of Reiten [63] (and independently of Foxby [23]) that $A \ltimes M$ is a Gorenstein local ring if and only if A is Cohen-Macaulay and M is a canonical A-module (in the sense of Herzog and Kunz [34]). We then employ the properties of a Gorenstein ring and the relations between $A \ltimes M$ and A and M to establish many of the properties of Gorenstein and canonical modules, first discovered by Sharp and Foxby. The game we play is this: If $A \ltimes M$ is a Gorenstein ring, then a property which is equivalent to being Gorenstein induces properties on both A and M (and conversely). One of the main tools is the natural isomorphism

$$\mathrm{Ext}^i_{A \ltimes M}(X, A \ltimes M) \cong \mathrm{Ext}^i_A(X, M \oplus \mathrm{Ann}_A M)$$

for all A-modules X, under the assumption $\mathrm{id}_{A \ltimes M}(A \ltimes M) < \infty$. This together with the "change of rings" theorems for regular sequences allows us to play the game very effectively. With the help of an example of Ferrand and Raynaud, we are able to show that not all Cohen-Macaulay local rings possess a Gorenstein module.

In section 6 we restrict our attention to algebras which are finite over a commutative artin ring. Such an algebra we call a finite algebra. The prototypes are finite dimensional algebras over a field. If R is a finite algebra, the dominant dimension of R is at least n, and we write dom.dim $R \geq n$, if in a minimal injective resolution $R \longrightarrow E^{\bullet}$ of the left R-module R, the modules E^i have flat dim $E^i = 0$ for $i < n$. Thus a finite algebra R with dom.dim $R \geq n$ is n-Gorenstein. In addition to constructing algebras with arbitrarily large dominant dimension, we study the relations between the category of reflexive finitely generated modules and finite algebras with dominant dimension at least 2.

In Section 7 we add our little result to Auslander's theory of representation dimension. Suppose Λ is a finite k-algebra. Let $\underline{\underline{A}}$ be a full additive subcategory of left Λ-modules generated by a finite number of indecomposable modules which contains all indecomposable projective and injective Λ-modules. Then $\underline{\underline{A}}$ is coherent and dom.dim $\mathrm{Coh}[\underline{\underline{A}}^{op}, \underline{\underline{Ab}}] \geq 2$, as is shown by Auslander in [4]. The representation dimension is defined by

$$\text{rep.dim}\,\Lambda = \inf_{\underline{\underline{A}}}\{\text{gl.dim}\,\mathrm{Coh}[\underline{\underline{A}}^{op}, \underline{\underline{Ab}}]\}.$$

If rep.dim $\Lambda \leq 2$, then Λ has a finite number of indecomposable finitely generated modules. One of the main results in this section is: rep.dim $T_2(\Lambda) \leq 2 +$ rep.dim Λ. Examples of Janusz and Brenner show that the representation dimension of $T_2(\Lambda)$ must grow (sometimes). Ringel has mentioned that rep.dim $T_2(T_2(T_2(T_2(\Lambda)))) > 2$ and this has been improved to rep.dim $T_2(T_2(T_2(\Lambda))) > 2$.

This is our last section.

We take this opportunity to thank our many colleagues for their stimulating and helpful discussions concerning this material. Especially helpful was Maurice Auslander who not only suggested problems, proofs and new ideas, but also has been constantly encouraging us. Others who deserve particular mention are Rodney Sharp, Hans-Bjørn Foxby, Lucien Szpiro, Gerald Janusz, and Birger Iversen, who have contributed ideas which have helped us to formulate these notions. The Mathematics departments of University of Illinois, Brandeis University and Aarhus Universitet have provided us with all the necessary resources for which we are grateful. We also acknowledge support from various other institutions. We have all received support from the United States National Science Foundation. Griffith has been supported by the Sloan Foundation and Reiten has received support from Norges Almenvitenskapelige Forskningsræd. Finally we thank Marcia Wolf and Janet Largent for preparing the camera-ready manuscript.

Robert Fossum
Phillip Griffith
Idun Reiten

Section 0. Preliminaries

Although most of the categories considered in this paper will be abelian, there will arise the occasion when only an additive category will be used. By an additive category we mean one with finite sums (or products).

We will use the convention that the composition $\alpha\beta$ of two morphisms is first β, then α. (However in some cases where a morphism α is a homomorphism of a left R-module M, then the value of x in M under α will be denoted by $x\alpha$. Then the composition $\alpha\beta$ will mean first α, then β.) Although this convention seems to be out of step with modern category theoretical notation, it is still consistent with standard usage in ring theory.

If $\underline{\underline{A}}$ is a category and α is a morphism in $\underline{\underline{A}}$, then dom α denotes the domain of α, cod α denotes the codomain of α, ker α denotes the kernel of α and cok α denotes the cokernel of α. We will be inconsistent in our usage of the kernel and cokernel. Actually these are morphisms rather than objects. However we will use the terminology interchangably, the possibility of confusion being minimal.

If A,B,C and D are objects in an additive category, then a morphism $A \oplus B \longrightarrow C \oplus D$ will be described by a matrix

$$\begin{pmatrix} a & b \\ c & d \end{pmatrix} : A \oplus B \longrightarrow C \oplus D$$

where $a:A \longrightarrow C$, $b:B \longrightarrow C$, $c:A \longrightarrow D$ and $d:B \longrightarrow D$. Then composition is ordinary matrix multiplication.

If R is a ring, then $\underline{\underline{\mathrm{Mod}}}_R$ (resp.: ${}_R\underline{\underline{\mathrm{Mod}}}$) is the category of right (resp.: left) R-modules.

If R and S are rings, M a right R-module and left S-module and N a left R-module and a right S-module, then the ring of matrices $\begin{pmatrix} R & N \\ M & S \end{pmatrix}_{(0,0)}$ will be the ring whose additive substructure is the coordinate structure of the product set $\begin{pmatrix} R & N \\ M & S \end{pmatrix}$ and with product

$$\begin{pmatrix} r & n \\ m & s \end{pmatrix} \begin{pmatrix} r' & n' \\ m' & s' \end{pmatrix} = \begin{pmatrix} rr' & rn' + ns' \\ mr' + sm' & ss' \end{pmatrix} .$$

For the ring R and the R-bimodule M, the ring $R \ltimes M$ is the ring whose underlying additive abelian group is the direct sum $R \times M$ with multiplication given by $(r,m)(r',m') = (rr',mr' + rm')$. The notation $R \ltimes M$, which is asymmetric, is adopted from the similar notation for semi-direct product for groups. It is a combination of the product sign $\times$ with the normal subgroup sign $\triangleright$. It is used in order to distinguish the product in the category of rings from the trivial extension. Thus, for example, if Q is the total ring of quotients of a commutative ring R, then we can and will form two rings: the product $R \times Q$ and the trivial extension $R \ltimes Q$. These have the same additive structure, but are not at all alike as rings.

Finally if $\underline{\underline{A}}$ is an abelian category with enough projective objects (resp.: injective objects) then $\mathrm{pd}_{\underline{\underline{A}}} A$ (resp.: $\mathrm{id}_{\underline{\underline{A}}} A$) denotes the projective dimension (resp.: injective dimension) of the object A.

The <u>global dimension</u> is denoted by gl.dim $\underline{\underline{A}}$ and is $\sup\{\mathrm{pd}_{\underline{\underline{A}}} A\colon A \in \underline{\underline{A}}\}$ (resp.: $\sup\{\mathrm{id}_{\underline{\underline{A}}} A\colon A \in \underline{\underline{A}}\}$).

The <u>finitistic projective dimension</u> of $\underline{\underline{A}}$, denoted $\mathrm{FPD}(\underline{\underline{A}})$, is the integer (or ∞) $\sup\{\mathrm{pd}_{\underline{\underline{A}}} A\colon \mathrm{pd}_{\underline{\underline{A}}} A < \infty\}$.

If R is a ring and $\underline{\underline{A}} = {}_R\underline{\underline{\mathrm{Mod}}}$, then $\mathrm{fPD}(R) = \sup\{\mathrm{pd}_{\underline{\underline{A}}} A\colon \mathrm{pd}_{\underline{\underline{A}}} A < \infty$ and A is finitely generated$\}$.

Section 1. Generalities

Let $\underline{\underline{A}}$ be an abelian category and $F: \underline{\underline{A}} \longrightarrow \underline{\underline{A}}$ an additive endofunctor. We describe new additive categories $F \rtimes \underline{\underline{A}}$ and $A \rtimes F$ by describing their objects, morphisms and composition.

We define first the right trivial extension of $\underline{\underline{A}}$ by F, denoted by $\underline{\underline{A}} \ltimes F$. An object in $\underline{\underline{A}} \ltimes F$ is a morphism $\alpha: FA \longrightarrow A$ for an object A in $\underline{\underline{A}}$ such that the composition $\alpha \cdot F\alpha = 0$. If $\alpha: FA \longrightarrow A$ and $\beta: FB \longrightarrow B$ are objects in $\underline{\underline{A}} \ltimes F$, then a morphism $\gamma: \alpha \longrightarrow \beta$ is a morphism $\gamma: A \longrightarrow B$ such that the diagram

$$\begin{array}{ccc} FA & \xrightarrow{F\gamma} & FB \\ {\scriptstyle\alpha}\downarrow & & \downarrow{\scriptstyle\beta} \\ A & \xrightarrow{\gamma} & B \end{array}$$

is commutative. Composition in $\underline{\underline{A}} \ltimes F$ is just composition in $\underline{\underline{A}}$.

The left trivial extension of $\underline{\underline{A}}$ by F, denoted by $F \rtimes \underline{\underline{A}}$, is the category whose objects are morphisms $\alpha: A \longrightarrow FA$ such that the composition $A \xrightarrow{\alpha} FA \xrightarrow{F\alpha} F^2A$ is zero. It has a description for morphisms similar to that for $\underline{\underline{A}} \ltimes F$ and composition is composition in $\underline{\underline{A}}$.

It is immediately clear that both $\underline{\underline{A}} \ltimes F$ and $F \rtimes \underline{\underline{A}}$ are additive categories. Furthermore, the category $\underline{\underline{A}} \ltimes F$ has kernels and $F \rtimes \underline{\underline{A}}$ has cokernels.

Proposition 1.1. a) *If F is right exact, then $\underline{\underline{A}} \ltimes F$ is abelian.*

b) *If F is left exact, then $F \rtimes \underline{\underline{A}}$ is abelian.*

Proof. We demonstrate b), the demonstration for a) being dual. To show that $F \rtimes \underline{\underline{A}}$ is abelian we must construct kernels and cokernels and show that monomorphisms are kernels and epimorphisms are cokernels [Freyd; 25]. Suppose $\alpha: A \to FA$ and $\beta: B \to FB$ are objects in $F \rtimes \underline{\underline{A}}$ and that $\gamma: \alpha \to \beta$ is a morphism. Let $\delta: B \to C$ be the cokernel of γ, when considered as a morphism in $\underline{\underline{A}}$. Then the diagram

$$\begin{array}{ccccccc} A & \xrightarrow{\gamma} & B & \xrightarrow{\delta} & C & \longrightarrow & 0 \\ \alpha\downarrow & & \beta\downarrow & & \downarrow\epsilon & & \\ FA & \xrightarrow{F\gamma} & FB & \xrightarrow{F\delta} & FC & & \end{array}$$

is commutative. Now the composition $F\delta \cdot \beta \cdot \gamma = F\delta \cdot F\gamma \cdot \alpha = 0$. Hence there is a unique $\epsilon: C \to FC$ such that $\epsilon \cdot \delta = F\delta \cdot \beta$. We need to show that $F\epsilon \cdot \epsilon = 0$. Since δ is an epimorphism, it is sufficient to show that $F\epsilon \cdot \epsilon \cdot \delta = 0$. But $F\epsilon \cdot \epsilon \cdot \delta = F\epsilon \cdot F\delta \cdot \beta = F(\epsilon \cdot \delta) \cdot \beta$. Furthermore $F(\epsilon \cdot \delta) = F(F\delta \cdot \beta) = F^2\delta \cdot F\beta$. Hence, $F\epsilon \cdot \epsilon \cdot \delta = F^2\delta \cdot F\beta \cdot \beta$. But $F\beta \cdot \beta = 0$, so $F\epsilon \cdot \epsilon = 0$.

Thus $\delta: \beta \to \epsilon$ is a candidate for the cokernel for γ in $F \rtimes \underline{\underline{A}}$. Suppose $\mu: \beta \to \eta$ is such that $\mu\gamma = 0$. Then the composition $A \xrightarrow{\gamma} B \xrightarrow{\eta} \operatorname{dom} \eta$ is zero, so there is a unique $\rho: C \to \operatorname{dom} \eta$ such that $\mu = \rho\delta$. We claim that the diagram

$$\begin{array}{ccc} C & \xrightarrow{\rho} & \operatorname{dom} \eta \\ \downarrow\epsilon & & \downarrow\eta \\ FC & \xrightarrow{F\rho} & F(\operatorname{dom} \eta) \end{array}$$

is commutative. Since δ is an epimorphism, it is enough to show that $(\eta \cdot \rho - F\rho \cdot \epsilon) \cdot \delta = 0$. However $\eta \cdot \rho \cdot \delta = \eta \cdot \mu = F\mu \cdot \beta$ while $F\rho \cdot \epsilon \cdot \delta = F\rho \cdot F\delta \cdot \beta = F(\rho \cdot \delta) \cdot \beta = F\mu \cdot \beta$. So $\rho: \epsilon \to \eta$ is a morphism in $F \rtimes \underline{\underline{A}}$. Therefore $\delta: \beta \to \epsilon$ is the cokernel of γ.

Now we use the left exactness of F to demonstrate that kernels exist. Let $K \xrightarrow{\kappa} A$ be the kernel, in $\underline{\underline{A}}$, of $\gamma: A \to B$. Then $0 \to K \xrightarrow{\kappa} A \to B$ is exact. Since F is left exact, we get a commutative diagram

$$\begin{array}{ccccccc} 0 & \longrightarrow & K & \xrightarrow{\kappa} & A & \xrightarrow{\gamma} & B \\ & & \epsilon'\downarrow & & \downarrow\alpha & & \downarrow\beta \\ 0 & \longrightarrow & FK & \xrightarrow{F\kappa} & FA & \xrightarrow{F\gamma} & FB \end{array}$$

with exact rows. Now $F\gamma \cdot \alpha \cdot \kappa = \beta \cdot \gamma \cdot \kappa = 0$, so there is a unique

$\epsilon'\colon K \longrightarrow FK$ such that $\alpha \cdot \kappa = F\kappa \cdot \epsilon'$, since $FK \xrightarrow{F\kappa} FA$ is the kernel of $F\gamma$.

Now we add a third row to our diagram which is also exact:

$$
\begin{array}{ccccccc}
0 & \longrightarrow & K & \xrightarrow{\kappa} & A & \xrightarrow{\gamma} & B \\
 & & \downarrow{\scriptstyle \epsilon'} & & \downarrow{\scriptstyle \alpha} & & \downarrow{\scriptstyle \beta} \\
0 & \longrightarrow & F\kappa & \xrightarrow{F\kappa} & FA & \xrightarrow{F\gamma} & FB \\
 & & \downarrow{\scriptstyle F\epsilon'} & & \downarrow{\scriptstyle F\alpha} & & \downarrow{\scriptstyle F\beta} \\
0 & \longrightarrow & F^2K & \xrightarrow{F^2\kappa} & F^2A & \xrightarrow{F^2\gamma} & F^2B
\end{array}
\quad .
$$

We need to show that $F\epsilon' \cdot \epsilon' = 0$. Since $F^2\kappa$ is a monomorphism, it is enough to show that $F^2\kappa \cdot F\epsilon' \cdot \epsilon' = 0$. But $F^2\kappa \cdot F\epsilon' \cdot \epsilon' = F\alpha \cdot \alpha \cdot \kappa = 0$. Now, as above, $\kappa\colon \epsilon' \longrightarrow \alpha$ is the kernel of $\gamma\colon \alpha \longrightarrow \beta$. Thus we have demonstrated the existence of kernels when F is left exact.

The morphism $\gamma\colon \alpha \longrightarrow \beta$ is a monomorphism (resp.: epimorphism) if and only if $\gamma\colon A \longrightarrow B$ is a monomorphism (resp.: epimorphism). For suppose $\gamma\colon \alpha \longrightarrow \beta$ is a monomorphism in $F \rtimes \underline{\underline{A}}$ and suppose the composition $K' \xrightarrow{\sigma} A \xrightarrow{\gamma} B$ is zero in $\underline{\underline{A}}$. We have just shown that $\kappa\colon \epsilon' \longrightarrow \alpha$ is the kernel of γ, where $K \xrightarrow{\kappa} A$ is the kernel of $A \xrightarrow{\alpha} B$. Since $\gamma\colon \alpha \longrightarrow \beta$ is a monomorphism, its kernel in $F \rtimes \underline{\underline{A}}$ is zero, so $\ker(A \xrightarrow{\gamma} B) = 0$. Hence $\sigma = 0$.

Now suppose $\gamma\colon \alpha \longrightarrow \beta$ is a monomorphism in $F \rtimes \underline{\underline{A}}$. It has a cokernel $\delta\colon \beta \longrightarrow \epsilon$, with $B \xrightarrow{\delta} C$ the cokernel of $\gamma\colon A \longrightarrow B$. Since $\underline{\underline{A}}$ is an abelian category, the map $\gamma\colon A \longrightarrow B$ is the kernel of $\delta\colon B \longrightarrow C$. Hence $\gamma\colon \alpha \longrightarrow \beta$ is the kernel of $\delta\colon \beta \longrightarrow \epsilon$.
Q.E.D.

We note a consequence which can be obtained from the proof.

Corollary 1.2. Suppose F *is right exact* (resp.: *left exact*). *Then a sequence of objects in* $\underline{\underline{A}} \ltimes F$ (resp.: $F \rtimes \underline{\underline{A}}$) *is exact if and only if the sequence of codomains* (resp.: *domains*) *is exact*. (Harada [32] has results similar to the first proposition.)

Before continuing with further results concerning the trivial extension categories, a few examples are given.

Suppose R is a ring (with 1) and M is an R-bimodule. Let $\underline{\underline{A}}$ denote the category of left R-modules. Then there are two "natural" functors associated with M, the tensor product $F = M \otimes_A -$ and the internal hom, $G = \mathrm{Hom}_A(M,-)$. We can also define the trivial extension of R by M to be the ring whose additive group is the direct sum $R \times M$ with multiplication given by $(r,m) \cdot (r',m') = (rr', mr' + rm')$. Denote this ring by $R \ltimes M$ (or $M \rtimes R$). We will see later that the categories $G \rtimes \underline{\underline{A}}$, $\underline{\underline{A}} \ltimes F$ and ${}_{R \ltimes M}\underline{\underline{\mathrm{Mod}}}$ are all isomorphic.

A more general example, one we will be continually using, is the comma category. Suppose $\underline{\underline{A}}$ and $\underline{\underline{B}}$ are abelian categories and $F\colon \underline{\underline{A}} \longrightarrow \underline{\underline{B}}$ is an additive functor. The comma category $(F,\underline{\underline{B}})$ is the category whose objects are triples (A,f,B) where $f\colon FA \longrightarrow B$ and whose morphisms are pairs (α,β) of morphisms in $\underline{\underline{A}} \times \underline{\underline{B}}$ such that the diagram

$$\begin{array}{ccc} FA & \xrightarrow{F\alpha} & FA' \\ \downarrow f & & \downarrow f' \\ B & \xrightarrow{\beta} & B' \end{array}$$

is commutative.

The functor F induces a functor $\tilde{F}\colon \underline{\underline{A}} \times \underline{\underline{B}} \longrightarrow \underline{\underline{A}} \times \underline{\underline{B}}$ by $\tilde{F}(A,B) = (0,FA)$ and $\tilde{F}(\alpha,\beta) = (0,F\alpha)$. It is not difficult to show that the categories $(F,\underline{\underline{B}})$ and $(\underline{\underline{A}} \times \underline{\underline{B}}) \ltimes \tilde{F}$ are isomorphic. For in fact a triple (A,f,B) is just the object $(0,f)\colon \tilde{F}(A,B) \longrightarrow (A,B)$. On the other hand, since $\tilde{F}^2 = 0$, any morphism $\tilde{F}(A,B) \longrightarrow (A,B)$ is an object in $(\underline{\underline{A}} \times \underline{\underline{B}}) \ltimes \tilde{F}$. But such a morphism is just a morphism $FA \longrightarrow B$. We leave to the reader the comparison for the morphisms. The category $(F,\underline{\underline{B}})$ or $(\underline{\underline{A}} \times \underline{\underline{B}}) \ltimes \tilde{F}$ will be denoted by $\underline{\underline{\mathrm{Map}}}(F\underline{\underline{A}},\underline{\underline{B}})$ in this paper.

We could also consider the comma category $(\underline{\underline{B}},F)$. An object in $(\underline{\underline{B}},F)$ is a triple (B,f,A) where $f\colon B \longrightarrow FA$. The morphisms are appropriately defined. Let $\tilde{F}\colon \underline{\underline{B}} \times \underline{\underline{A}} \longrightarrow \underline{\underline{B}} \times \underline{\underline{A}}$ be defined by $\tilde{F}(B,A) = (FA,0)$. Now an object in $\tilde{F} \rtimes (\underline{\underline{B}} \times \underline{\underline{A}})$ is a morphism $(B,A) \longrightarrow (FA,0)$; just an element in $(\underline{\underline{B}},F)$.

Thus, if F is right exact and so $\tilde{F}\colon \underline{\underline{A}} \times \underline{\underline{B}} \longrightarrow \underline{\underline{A}} \times \underline{\underline{B}}$ is right exact, then the category $\underline{\underline{\mathrm{Map}}}(F\underline{\underline{A}},\underline{\underline{B}})$ is abelian. Likewise, if F is left exact and so $\tilde{F}\colon \underline{\underline{B}} \times \underline{\underline{A}} \longrightarrow \underline{\underline{B}} \times \underline{\underline{A}}$ is left exact, then the category $F \rtimes (\underline{\underline{B}} \times \underline{\underline{A}})$, which we will denote by $\underline{\underline{\mathrm{Map}}}(\underline{\underline{B}},F\underline{\underline{A}})$, is abelian.

From this point to the end of the paper, unless mentioned to the contrary, whenever the category $\underline{\underline{A}} \ltimes F$ is considered, it will be assumed that F is right exact. A corresponding assumption will be made so that the categories $F \rtimes \underline{\underline{A}}$, $\underline{\underline{\mathrm{Map}}}(F\underline{\underline{A}},\underline{\underline{B}})$ and $\underline{\underline{\mathrm{Map}}}(\underline{\underline{B}},F\underline{\underline{A}})$ will be abelian.

For each endofunctor $F\colon \underline{\underline{A}} \longrightarrow \underline{\underline{A}}$ there are pairs of adjoint functors

$$\underline{\underline{A}} \underset{U}{\overset{T}{\rightleftarrows}} \underline{\underline{A}} \ltimes F \underset{Z}{\overset{C}{\rightleftarrows}} \underline{\underline{A}}$$

and pairs

$$\underline{\underline{A}} \underset{K}{\overset{Z}{\rightleftarrows}} F \rtimes \underline{\underline{A}} \underset{H}{\overset{U}{\rightleftarrows}} \underline{\underline{A}} \quad .$$

These satisfy the relations:

$$CT = \mathrm{id}_{\underline{\underline{A}}}$$

$$UZ = \mathrm{id}_{\underline{\underline{A}}}$$

and

$$KH = \mathrm{id}_{\underline{\underline{A}}} \quad .$$

They are defined on objects and morphisms as follows:

The <u>tensor</u> $T\colon \underline{\underline{A}} \longrightarrow \underline{\underline{A}} \ltimes F$ is defined on objects by

$$T(A) = \begin{pmatrix} 0 & 0 \\ 1 & 0 \end{pmatrix}\colon \; FA \oplus F^2A \longrightarrow A \oplus FA$$

and on morphisms by

$$T(\alpha) = \begin{pmatrix} \alpha & 0 \\ 0 & F\alpha \end{pmatrix}.$$

The <u>underlying</u> functor $U\colon \underline{\underline{A}} \ltimes F \longrightarrow \underline{\underline{A}}$ is defined by $U(FA \xrightarrow{\alpha} A) = \operatorname{codom} \alpha = A$ and $U(\alpha) = \alpha$.

The zero functor $Z: \underline{\underline{A}} \longrightarrow \underline{\underline{A}} \ltimes F$ is given by $Z(A) = 0:$ $FA \longrightarrow A$ and $Z(\alpha) = \alpha$.

The cokernel functor $C: \underline{\underline{A}} \ltimes F \longrightarrow \underline{\underline{A}}$ is defined by $C(FA \xrightarrow{\alpha} A) = \text{coker}\ \alpha$ while $C(\gamma)$ is the induced map.

Dually we define the hom functor $H: \underline{\underline{A}} \longrightarrow F \rtimes \underline{\underline{A}}$ by

$H(A) = \begin{pmatrix} 0 & 0 \\ 1 & 0 \end{pmatrix}: \ FA \oplus A \longrightarrow F^2A \oplus FA$ and $H(\alpha) = \begin{pmatrix} F\alpha & 0 \\ 0 & \alpha \end{pmatrix}$.

The underlying functor U is now the domain.

The zero functor is again the zero map while the kernel functor K is the kernel on objects and the induced map on morphisms.

Proposition 1.3. *The functor* T *is left adjoint to* U *and* C *is left adjoint to* Z. *The functor* U *is left adjoint to* H *and* Z *is left adjoint to* K.

Proof. We demonstrate just one pair of adjointness relations. We show $T \dashv U$. We must show that $\text{Hom}_{\underline{\underline{A}} \ltimes F}(TA,\beta) \cong \text{Hom}_{\underline{\underline{A}}}(A,U\beta)$ naturally in A and β. Suppose $(\gamma,\delta): A \oplus FA \longrightarrow B$ is a morphism in $\underline{\underline{A}}$ which is also in $\underline{\underline{A}} \ltimes F$. Then $\beta \cdot F(\gamma,\delta) = (\gamma,\delta)\begin{pmatrix} 0 & 0 \\ 1 & 0 \end{pmatrix}$. Now $(\gamma,\delta)\begin{pmatrix} 0 & 0 \\ 1 & 0 \end{pmatrix} = (\delta,0)$. Hence we have $(\beta \cdot F\gamma, \beta \cdot F\delta) = (\delta,0)$. Thus $\delta = \beta \cdot F\gamma$. If $\delta = \beta \cdot F\gamma$, then $\beta \cdot F\delta = \beta \cdot F\beta \cdot F^2\delta = 0 \cdot F^2\delta = 0$. Hence, the second coordinates are equal if the first are. Then $(\gamma,\delta) \in \underline{\underline{A}} \times F$ if and only if $(\gamma,\delta) = (\gamma, \beta \cdot F\gamma)$. So define $\text{Hom}_{\underline{\underline{A}} \ltimes F}(TA,\beta) \longrightarrow \text{Hom}_{\underline{\underline{A}}}(A,U\beta)$ by $(\gamma, \beta \cdot F\delta) \longmapsto \gamma$. The obvious inverse is the map $\text{Hom}_{\underline{\underline{A}}}(A,U\beta) \longrightarrow \text{Hom}_{\underline{\underline{A}} \ltimes F}(TA,\beta)$ by $\gamma \longmapsto (\gamma, \beta \cdot F\gamma)$. These are clearly natural in A and β. Q.E.D.

For possible future reference, we record the other adjointness isomorphisms.

$$\text{Hom}_{\underline{\underline{A}} \ltimes F}(TA,\beta) \cong \text{Hom}_{\underline{\underline{A}}}(A,U\beta)$$

$$\text{Hom}_{\underline{\underline{A}}}(C\beta,B) \cong \text{Hom}_{\underline{\underline{A}} \ltimes F}(\beta,ZB),$$

$$\text{Hom}_{F \rtimes \underline{\underline{A}}}(ZA,\beta) \cong \text{Hom}_{\underline{\underline{A}}}(A,K\beta),$$

and

$$\text{Hom}_{\underline{\underline{A}}}(U\alpha,B) \cong \text{Hom}_{F \rtimes \underline{\underline{A}}}(\alpha,HB).$$

Lemma 1.4. *The object* $\alpha = 0$ *in* $\underline{\underline{A}} \ltimes F$ *if and only if coker* $\alpha = 0$. *The object* $\alpha = 0$ *in* $F \rtimes \underline{\underline{A}}$ *if and only if* $\ker \alpha = 0$. (N.B. We mean by $\alpha = 0$, that α is the zero object $0\colon FO \longrightarrow 0$.)

Proof. Suppose coker $\alpha = 0$. Then α is an epimorphism and so $F\alpha$ is an epimorphism. But $\alpha \cdot F\alpha = 0$, so $\alpha = 0$. Hence coker α = codom $\alpha = 0$.

QED.

Proposition 1.5. a) *The left adjunction* $\mathrm{id}_{\underline{\underline{A}} \ltimes F} \longrightarrow ZC$ *is a minimal epimorphism, in the sense that, for all* α *in* $\underline{\underline{A}} \ltimes F$, *the morphism* $\alpha \longrightarrow ZC\alpha$ *is a minimal epimorphism.*

b) *The right adjunction* $ZK \longrightarrow \mathrm{id}_{F \rtimes \underline{\underline{A}}}$ *is an essential monomorphism.*

Proof. Recall that a morphism $A \xrightarrow{\alpha} B$ is a minimal epimorphism if i) it is an epimorphism and ii) given any nonzero epimorphism $\delta\colon A \longrightarrow C$, the pushout of

$$\begin{array}{ccc} A & \xrightarrow{\alpha} & B \\ {\scriptstyle\delta}\downarrow & & \\ C & & \end{array}$$

is not zero. (Compare this with the usual definition of an essential monomorphism.)

Let $\alpha\colon FA \longrightarrow A$ be an object in $\underline{\underline{A}} \ltimes F$. Let C denote its cokernel. We get the commutative diagram

$$\begin{array}{ccc} FA & \xrightarrow{F\pi} & FC \\ \downarrow{\scriptstyle\alpha} & & \downarrow{\scriptstyle 0} \\ A & \xrightarrow{\pi} & C \end{array}$$

which is an epimorphism, since $A \xrightarrow{\pi} C$ is an epimorphism. Suppose

$$\begin{array}{ccc} FA & \xrightarrow{F\delta} & FD \\ \downarrow{\scriptstyle\alpha} & & \downarrow{\scriptstyle\gamma} \\ A & \xrightarrow{\delta} & D \end{array}$$

is another nonzero epimorphism. Then the composition

$$\begin{array}{ccccc} FA & \xrightarrow{F\delta} & FD & \xrightarrow{F\rho} & FC(\gamma) \\ \downarrow \alpha & & \downarrow \gamma & & \downarrow 0 \\ A & \xrightarrow{\delta} & D & \xrightarrow{\rho} & C(\gamma) \end{array}$$

is also an epimorphism which is not zero by the previous lemma. But then $\rho \cdot \delta \cdot \alpha = 0$. Hence there is a unique σ: $C \longrightarrow C(\gamma)$ in $\underline{\underline{A}}$ such that $\rho \cdot \delta = \sigma \cdot \pi$, or that the diagram

$$\begin{array}{ccc} A & \xrightarrow{\pi} & C \\ \downarrow \delta & & \downarrow \sigma \\ D & \xrightarrow{\rho} & C(\gamma) \end{array}$$

commutes. But then there is a nonzero pushout. QED.

We record some consequences of the last two propositions.

Corollary 1.6. a) The functor T is right exact and U is exact. The functor Z is exact and C is right exact.

b) The functors Z and U are exact and K and H are left exact.

c) If P is projective in $\underline{\underline{A}}$ (resp.: $\underline{\underline{A}} \ltimes F$), then $T(P)$ (resp.: $C(P)$) is projective in $\underline{\underline{A}} \ltimes F$ (resp.: $\underline{\underline{A}}$). Consequently π is projective in $\underline{\underline{A}} \ltimes F$ if and only if $C(\pi)$ is projective in $\underline{\underline{A}}$ and $\pi \cong T(C(\pi))$.

d) If E is injective in $\underline{\underline{A}}$ (resp.: $F \rtimes \underline{\underline{A}}$), then $H(E)$ (resp.: KE) is injective in $F \rtimes \underline{\underline{A}}$ (resp.: $\underline{\underline{A}}$). Consequently ϵ in $F \rtimes \underline{\underline{A}}$ is injective if and only if $K\epsilon$ is injective and $\epsilon \cong H(K\epsilon)$.

Proof. Left adjoints to exact functors are right exact while right adjoints to exact functors are left exact. That Z and U are exact follows from Corollary 1.2.

Suppose P is projective in $\underline{\underline{A}}$ and that $\alpha \longrightarrow \alpha''$ is an epimorphism in $\underline{\underline{A}} \ltimes F$. Then we have

$$\begin{array}{ccc} \mathrm{Hom}_{\underline{\underline{A}} \ltimes F}(TP,\alpha) & \longrightarrow & \mathrm{Hom}_{\underline{\underline{A}} \ltimes F}(TP,\alpha'') \\ \downarrow \cong & & \downarrow \cong \\ \mathrm{Hom}_{\underline{\underline{A}}}(P,U\alpha) & \longrightarrow & \mathrm{Hom}_{\underline{\underline{A}}}(P,U\alpha''). \end{array}$$

But $U\alpha \longrightarrow U\alpha''$ is an epimorphism. Since P is projective in $\underline{\underline{A}}$, the homomorphism

$$\mathrm{Hom}_{\underline{\underline{A}} \ltimes F}(TP, \alpha) \longrightarrow \mathrm{Hom}_{\underline{\underline{A}} \ltimes F}(TP, \alpha'')$$

is surjective.

A more general statement is: Functors which are left adjoint to exact functors preserve projectives, while right adjoints to exact functors preserve injectives.

Suppose π in $\underline{\underline{A}} \ltimes F$ is projective. Then the object $C\pi$ is projective in $\underline{\underline{A}}$. The two morphisms $\pi \longrightarrow ZC\pi$ and $T(C\pi) \longrightarrow ZC\pi$ are minimal epimorphisms. Therefore any morphism $\pi \longrightarrow T(C\pi)$ such that the diagram

$$\begin{array}{ccc} & \swarrow & \pi \\ & & \downarrow \\ T(C\pi) & \longrightarrow & ZC\pi \end{array}$$

is commutative, is an isomorphism. QED.

Corollary 1.6 gives us a complete description of the projective objects in $\underline{\underline{A}} \ltimes F$ in terms of projective objects in $\underline{\underline{A}}$, while it gives a description of the injectives in $F \rtimes \underline{\underline{A}}$. Moreover, if injective envelopes exist, or projective covers exist in $\underline{\underline{A}}$, then they also exist in $F \rtimes \underline{\underline{A}}$ (resp.: $\underline{\underline{A}} \ltimes F$).

Corollary 1.7. a) *Suppose* $\underline{\underline{A}}$ *has projective covers. Then* $\underline{\underline{A}} \ltimes F$ *has projective covers. If* α: $FA \longrightarrow A$ *is an object in* $\underline{\underline{A}} \ltimes F$ *and* $P \longrightarrow C(\alpha)$ *is a projective cover in* $\underline{\underline{A}}$, *then* $T(P) \longrightarrow \alpha$ *is a projective cover in* $\underline{\underline{A}} \ltimes F$.

b) *Suppose* $\underline{\underline{A}}$ *has injective envelopes. Then* $F \rtimes \underline{\underline{A}}$ *has injective envelopes. If* β: $A \longrightarrow FA$ *is an object in* $F \rtimes \underline{\underline{A}}$ *and* $E(K(\beta))$ *is an injective envelope of* $K(\beta)$, *then an injective envelope of* α *is* $H(E(K(\beta)))$.

Proof. We know that $T(P) \longrightarrow ZP$ is a minimal epimorphism. Also $ZP \longrightarrow ZC(\alpha)$ is a minimal epimorphism. Hence there is a morphism $T(P) \longrightarrow \alpha$ such that

$$\begin{array}{ccc} & & T(P) \\ & \swarrow & \downarrow \\ \alpha & \longrightarrow & ZC(\alpha) \end{array}$$

is commutative. This must be a minimal epimorphism. QED.

The next result gives a partial explanation of the relation between the construction $\underline{\underline{A}} \ltimes F$ and the rings $R \ltimes M$.

<u>Proposition</u> 1.8. <u>Let</u> A <u>be an object in</u> $\underline{\underline{A}}$: <u>Then there is an isomorphism</u>

$$\mathrm{End}_{\underline{\underline{A}} \ltimes F}(T(A)) \cong \mathrm{End}_{\underline{\underline{A}}}(A) \ltimes \mathrm{Hom}_{\underline{\underline{A}}}(A,FA).$$

(And also we have

$$\mathrm{End}_{F \rtimes \underline{\underline{A}}}(H(A)) \cong \mathrm{End}_{\underline{\underline{A}}}(A) \ltimes \mathrm{Hom}_{\underline{\underline{A}}}(FA,A).)$$

Proof. We have

$$\mathrm{Hom}_{\underline{\underline{A}} \ltimes F}(TA,TB) \cong \mathrm{Hom}_{\underline{\underline{A}}}(A,UTB) \cong \mathrm{Hom}_{\underline{\underline{A}}}(A,B \oplus FB).$$

by the adjointness and the definition of T. Hence

$$\mathrm{End}_{\underline{\underline{A}} \ltimes F}(TA) \cong \mathrm{Hom}_{\underline{\underline{A}}}(A,A \oplus FA) \cong \mathrm{End}_{\underline{\underline{A}}}(A) \times \mathrm{Hom}_{\underline{\underline{A}}}(A,FA).$$

We need to show that the multiplication is that of the trivial ring extension. Well, any morphism is of the form $\begin{pmatrix} a & 0 \\ x & Fa \end{pmatrix}: A \oplus FA \longrightarrow A \oplus FA$, where $a \in \mathrm{End}_{\underline{\underline{A}}}(A)$ and $x \in \mathrm{Hom}_{\underline{\underline{A}}}(A,FA)$. This has as image in $\mathrm{End}_{\underline{\underline{A}}}(A) \times \mathrm{Hom}_{\underline{\underline{A}}}(A,FA)$ the element (a,x). Now

$$\begin{pmatrix} a & 0 \\ x & Fa \end{pmatrix}\begin{pmatrix} b & 0 \\ y & Fb \end{pmatrix} = \begin{pmatrix} ab & 0 \\ x\cdot b+Fa\cdot y & F(ab) \end{pmatrix}$$

which has $(ab, xb+Fa\cdot y)$ as its image. The $\mathrm{End}_{\underline{\underline{A}}}(A)$-bimodule $\mathrm{Hom}_{\underline{\underline{A}}}(A,FA)$ is described by ordinary composition on the right:

$$\mathrm{Hom}_{\underline{\underline{A}}}(A,FA) \times \mathrm{End}_{\underline{\underline{A}}}(A) \longrightarrow \mathrm{Hom}_{\underline{\underline{A}}}(A,FA)$$

by $(x,a) \longmapsto xa$. On the left we have

$$\mathrm{End}_{\underline{\underline{A}}}(A) \times \mathrm{Hom}_{\underline{\underline{A}}}(A,FA) \longrightarrow \mathrm{End}_{\underline{\underline{A}}}(FA) \times \mathrm{Hom}_{\underline{\underline{A}}}(A,FA) \longrightarrow \mathrm{Hom}_{\underline{\underline{A}}}(A,FA)$$

given by $(a,x) \longmapsto Fa \cdot x$. Thus the multiplication in $\mathrm{End}_{\underline{\underline{A}}}(A) \times \mathrm{Hom}_{\underline{\underline{A}}}(A,FA)$ is that given for the trivial extension. QED.

We should remark:

$$\mathrm{Hom}_{\underline{\underline{A}} \ltimes F}(TA,TB) \cong \mathrm{Hom}_{\underline{\underline{A}}}(A,B) \ltimes \mathrm{Hom}_{\underline{\underline{A}}}(A,FB),$$

by $\begin{pmatrix} r & 0 \\ x & Fr \end{pmatrix} \longleftrightarrow \begin{pmatrix} r \\ x \end{pmatrix}$.

The reason for stating this proposition at this point is to study the automorphisms of a projective object $T(P)$ in $\underline{\underline{A}} \ltimes F$ (and likewise of an injective object in $F \rtimes \underline{\underline{A}}$).

If R is a ring and M is an R-bimodule, then we have a complete description of the units in $R \ltimes M$. Let S^* denote the group of units in a ring S.

If M is an R-bimodule and u is a unit in R, the map defined by $m \longmapsto u^{-1}mu = m^u$ is an automorphism of M. So we have induced a group homomorphism $R^* \longrightarrow \mathrm{Aut}_{\mathbb{Z}}(M)$. We can then form the semidirect product $R^* \ltimes M$ whose elements are pairs $[u,m]$, with multiplication given by $[u,m]\,[v,n] = [uv, m^v + n]$.

<u>Lemma</u> 1.9. <u>The group of units</u> $(R \ltimes M)^* \cong R^* \ltimes M$.

Proof. Suppose u is unit in R and $m \in M$. Then (u,m) is a unit in $R \ltimes M$ with inverse $(u^{-1}, -u^{-1}mu^{-1})$. Conversely, if (u,m) is a unit in $R \ltimes M$, then u is a unit in R. Define $(R \ltimes M)^* \longrightarrow R^* \times M$ by $(u,m) \longmapsto [u, u^{-1}m]$. Now $(u,m)(v,n) = (uv, mv + un)$. However $(uv, mv + un) \longmapsto [uv, v^{-1}(u^{-1}m)v + v^{-1}n]$ while $[u,u^{-1}m]\,[v,v^{-1}n] = [uv, (u^{-1}m)^v + v^{-1}n]$. So this map is a group homomorphism, which is then obviously an isomorphism. QED.

<u>Corollary</u> 1.10. <u>The group of automorphisms</u>

$$\mathrm{Aut}_{\underline{\underline{A}} \ltimes F}(T(A)) \cong \mathrm{Aut}_{\underline{\underline{A}}}(A) \ltimes \mathrm{Hom}_{\underline{\underline{A}}}(A,FA). \qquad \text{QED.}$$

When does there exist a nice description of injective objects in $\underline{\underline{A}} \ltimes F$? Using the same argument used in the proof of Proposition 1.6, we can state: If T is exact, then U preserves injective objects. The functor T is exact when and only when F is exact. However, it need not be the case that every injective object is of the form $T(E)$ for an injective E. Nor does it seem possible to be able to describe injectives in $\underline{\underline{A}} \ltimes F$ even if they exist. The same holds for projectives in $F \rtimes \underline{\underline{A}}$. However, there is a very general situation, which arises quite often, in which we can compute injectives in $\underline{\underline{A}} \ltimes F$ (and projectives in $F \rtimes \underline{\underline{A}}$).

Proposition 1.11. *Suppose F is left adjoint to G. Then the categories $\underline{\underline{A}} \ltimes F$ and $G \rtimes \underline{\underline{A}}$ are isomorphic.*

Proof. For each pair A,B in $\underline{\underline{A}}$, there is a natural isomorphism η: $\mathrm{Hom}_{\underline{\underline{A}}}(FA,B) \cong \mathrm{Hom}_{\underline{\underline{A}}}(A,GB)$. It is easy to show that $\alpha \cdot F\alpha = 0$ when and only when $G(\eta(\alpha)) \cdot \eta(\alpha) = 0$. Thus there is an isomorphism between the objects of $\underline{\underline{A}} \ltimes F$ and the objects of $G \rtimes \underline{\underline{A}}$. The rest of the proof is left for the reader. QED.

Let ρ: $FG \longrightarrow \mathrm{id}_{\underline{\underline{A}}}$ and λ: $\mathrm{id}_{\underline{\underline{A}}} \longrightarrow GF$ denote the right and left adjunctions respectively. The isomorphism η then has an explicit description. Suppose α: $FA \longrightarrow A$. Then we get a composition $A \xrightarrow{\lambda A} GFA \xrightarrow{G\alpha} GA$ which is $\eta(\alpha)$. Returning we get, for β: $A \longrightarrow GA$, a composition $FA \xrightarrow{F\beta} FGA \xrightarrow{\rho A} A$, which is $\eta^{-1}(\beta)$.

Suppose we have enough injective objects in $\underline{\underline{A}}$. Then there are enough injectives in $G \rtimes \underline{\underline{A}}$.

Let α: $FA \longrightarrow A$ be an object in $\underline{\underline{A}} \ltimes F$. Let β: $A \longrightarrow GA$ be its corresponding object in $G \rtimes \underline{\underline{A}}$. Let E be an injective envelope of $\mathrm{Ker}\,\beta$. Then an injective envelope of β is

$$\begin{array}{c} GE \oplus E \\ \Big\downarrow \begin{pmatrix} 0 & 0 \\ 1 & 0 \end{pmatrix} \\ G^2E \oplus GE \end{array}$$

Perhaps it is wise to show the existence of a monomorphism. Since $0 \longrightarrow \ker\beta \longrightarrow A$ is a monomorphism, there is a morphism ι: $A \longrightarrow E$ such that

$$\begin{array}{ccc} \ker\beta & \longrightarrow & A \\ & \searrow \quad \swarrow \scriptstyle{\iota} & \\ & E & \end{array}$$

is commutative. Then $\binom{G\iota \;\cdot\; \beta}{\iota} : B \longrightarrow GE \oplus E$ is a monomorphism which is essential when considered in $G \rtimes \underline{\underline{A}}$.

Now the object in $\underline{\underline{A}} \ltimes F$ corresponding to

$$H(E) = \begin{array}{c} GE \oplus E \\ \downarrow \begin{pmatrix} 0 & 0 \\ 1 & 0 \end{pmatrix} \\ G^2E \oplus GE \end{array} \quad \text{is the object} \quad \begin{array}{c} FGE \oplus FE \\ \downarrow \begin{pmatrix} 0 & 0 \\ \rho & 0 \end{pmatrix} \\ GE \oplus E \end{array} .$$

This is injective and essential over $\alpha\colon FA \longrightarrow A$, so is the injective envelope of α in $\underline{\underline{A}} \ltimes F$.

Before treating some special cases, one more general result is given.

Proposition 1.12. a) *Suppose* $\underline{\underline{A}}$ *has enough projectives. Then there is an isomorphism of functors* $L_iT \cong ZL_iF$ *for all* $i > 0$ (where L_i is the ith derived functor).

b) *Suppose* $\underline{\underline{A}}$ *has enough injectives. There is an isomorphism of functors* $R^iH \cong ZR^iF$ *for all* $i > 0$.

Proof. There is an exact sequence of functors

$$0 \longrightarrow ZF \longrightarrow T \longrightarrow Z \longrightarrow 0$$

from $\underline{\underline{A}}$ to $\underline{\underline{A}} \ltimes F$. Since Z is exact, its left derived functors vanish. Hence the natural transformation $L_iZF \longrightarrow L_iT$ is an isomorphism for $i > 0$. But $L_iZF = ZL_iF$ since Z is exact. QED.

Now we state the preceding results for the categories $\underline{\text{Map}}(F\underline{\underline{A}},\underline{\underline{B}})$ and $\underline{\text{Map}}(\underline{\underline{A}},G\underline{\underline{B}})$ (where we assume that F is right exact, that G is left exact and, in some cases, F is left adjoint to G). The main reason for this is that in section 4 we will compute various homological dimensions which can be done rather easily for the comma categories, but which are much more difficult for the trivial extension categories. In these computations, the calculus of projective and injective objects is necessary.

Recall that $\underline{\text{Map}}(F\underline{\underline{A}},\underline{\underline{B}}) = (\underline{\underline{A}} \times \underline{\underline{B}}) \ltimes \tilde{F}$, where $\tilde{F}(A,B) = (0,FA)$. The functor $T\colon \underline{\underline{A}} \times \underline{\underline{B}} \longrightarrow (\underline{\underline{A}} \times \underline{\underline{B}}) \ltimes \tilde{F}$ then takes the form:

$$T(A,B) = \begin{array}{c} \tilde{F}(A,B) \oplus \tilde{F}^2(A,B) \\ \downarrow \begin{pmatrix} 0 & 0 \\ 1 & 0 \end{pmatrix} \\ (A,B) \oplus \tilde{F}(A,B) \end{array} = \begin{array}{c} (0,FA) \\ \downarrow = \\ (A,B \oplus FA). \end{array}$$

The zero functor:
$$Z(A,B) = \begin{array}{c} (0,FA) \\ \downarrow 0 \\ (A,B) \end{array}$$

For $\tilde{G} \rtimes (\underline{\underline{B}} \times \underline{\underline{A}})$, we have $H\colon \underline{\underline{B}} \times \underline{\underline{A}} \longrightarrow \tilde{G} \rtimes (\underline{\underline{B}} \times \underline{\underline{A}})$ given by:

$$H(B,A) = \begin{array}{c} \tilde{G}\,(B,A) \oplus (B,A) \\ \downarrow \begin{pmatrix} 0 & 0 \\ 1 & 0 \end{pmatrix} \\ \tilde{G}^2(B,A) \oplus \tilde{G}(B,A) \end{array} = \begin{array}{c} (GA \oplus B,A) \\ \downarrow = \\ (GA,0) \end{array} .$$

The zero functor is:
$$Z(B,A) = \begin{array}{c} (B,A) \\ \downarrow 0 \\ (GA,0) \end{array}$$

The explicit descriptions of the remaining functors we leave for exercises.

An object (A,B) in $\underline{\underline{A}} \times \underline{\underline{B}}$ is projective if and only if A is projective in $\underline{\underline{A}}$ and B is projective in $\underline{\underline{B}}$. Thus, a projective object in $(\underline{\underline{A}} \times \underline{\underline{B}}) \ltimes \tilde{F}$ is isomorphic to $T(A,B)$ where A is projective in $\underline{\underline{A}}$ and B is projective in $\underline{\underline{B}}$. So a projective is a morphism

$$\begin{array}{c} (0, \quad FP) \\ \downarrow = \\ (P,Q \oplus FP) \end{array} \quad \begin{array}{l} P \text{ projective in } \underline{\underline{A}} \text{ and} \\ \\ Q \text{ projective in } \underline{\underline{B}}\ . \end{array}$$

Likewise an injective object in $\tilde{G} \rtimes (\underline{\underline{B}} \times \underline{\underline{A}})$ is of the form

$$\begin{array}{c} (GI \oplus J,I) \\ \downarrow = \\ (GI, \quad 0) \end{array}$$

for I injective in $\underline{\underline{A}}$ and J injective in $\underline{\underline{B}}$. The computation of the ring of endomorphisms of an extended object $T(A,B)$, from Proposition 1.8, leads to the isomorphism

$$\mathrm{End}_{(\underline{\underline{A}} \times \underline{\underline{B}}) \ltimes \tilde{F}}(T(A,B)) \cong \mathrm{End}_{\underline{\underline{A}} \times \underline{\underline{B}}}(A,B) \ltimes \mathrm{Hom}_{\underline{\underline{A}} \times \underline{\underline{B}}}((A,B),\tilde{F}(A,B)).$$

But $\mathrm{End}_{\underline{\underline{A}} \times \underline{\underline{B}}}(A,B) \quad = \quad \mathrm{End}_{\underline{\underline{A}}}(A) \times \mathrm{End}_{\underline{\underline{B}}}(B)$

while $\qquad \mathrm{Hom}_{\underline{\underline{A}} \times \underline{\underline{B}}}((A,B),\tilde{F}(A,B)) = \mathrm{Hom}_{\underline{\underline{B}}}(B,FA).$

Now $\mathrm{Hom}_{\underline{\underline{B}}}(B,FA)$ is a right $\mathrm{End}_{\underline{\underline{A}}}(A)$-module and a left $\mathrm{End}_{\underline{\underline{B}}}(B)$-module (according to our conventions). Thus

$$\mathrm{End}_{(\underline{\underline{A}} \times \underline{\underline{B}}) \ltimes \tilde{F}}(T(A,B)) \cong (\mathrm{End}_{\underline{\underline{A}}}(A) \times \mathrm{End}_{\underline{\underline{B}}}(B)) \ltimes \mathrm{Hom}_{\underline{\underline{B}}}(B,FA).$$

Suppose R and S are rings and M is a left S-module and right R-module. Then the ring of "matrices" $\begin{pmatrix} R & 0 \\ M & S \end{pmatrix}$, with "obvious" multiplication, is isomorphic to the trivial extension $(R \times S) \ltimes M$, where M attains a right $R \times S$ structure through the homomorphism $R \times S \longrightarrow R$ and a left $R \times S$ structure through the homomorphism $R \times S \longrightarrow S$. (For example, if A is a commutative ring with a symmetric module M, then M becomes an $A \times A$ bimodule which is not symmetric.) Thus we can write

$$\mathrm{End}_{(\underline{\underline{A}} \times \underline{\underline{B}}) \ltimes \tilde{F}}(T(A,B)) = \begin{pmatrix} \mathrm{End}_{\underline{\underline{A}}}(A) & 0 \\ \mathrm{Hom}_{\underline{\underline{B}}}(B,FA) & \mathrm{End}_{\underline{\underline{B}}}(B) \end{pmatrix}.$$

The derived functors of T and H, as computed in Proposition 1.12, take the following form: We have for $i > 0$

$$L_iT(A,B) \cong ZL_i\tilde{F}(A,B) \ .$$

by Proposition 1.12. However $(A,B) = (A,0) \oplus (0,B)$. So in order to compute $L_iT(A,B)$, it is enough to compute $L_iT(A,0)$ and $L_iT(0,B)$. But T is the identity on $0 \times \underline{\underline{B}}$, so $L_iT(0,B) = 0$ for $i > 0$. A projective resolution of $(A,0)$ is of the form $(P^{\cdot},0)$ where $P^{\cdot} \longrightarrow A$ is a projective resolution of A in $\underline{\underline{A}}$.

Now $\tilde{F}(P^\cdot,0) = (0,FP^\cdot)$, so $L_i\tilde{F}(A,0) = (0,L_iFA)$ for all i. Hence we conclude $L_iT(A,B) = Z(0,L_iFA)$ for $i > 0$. This is the object

$$\begin{array}{c}(0,0)\\ \downarrow \\ (0,L_iFA)\end{array}$$

in $(\underline{\underline{A}} \times \underline{\underline{B}}) \ltimes \tilde{F}$.

A similar statement for $\tilde{G} \rtimes (\underline{\underline{B}} \times \underline{\underline{A}})$ is:

$$R^iH(B,A) = \begin{array}{c}(0,R^iGA)\\ \downarrow \\ (0,0)\end{array}$$

in $\tilde{G} \rtimes (\underline{\underline{B}} \times \underline{\underline{A}})$.

We turn to a complete description of the relation between the trivial extension of a ring by a bimodule and the corresponding notion for a category. Let A be a ring and M an A-bimodule. Let B be the ring $A \ltimes M$ and $\iota\colon A \longrightarrow B$ the ring homomorphism to the first coordinate.

Suppose X is a left B-module. Then X is a left A-module through ι. Furthermore, the B-module structure of X is uniquely determined by its A-module structure together with a homomorphism of left A-modules $\mu\colon B \otimes_A X \longrightarrow X$ such that the diagrams

$$\begin{array}{ccc} X & \xrightarrow{\iota \otimes X} & B \otimes_A X \\ & {\scriptstyle =}\searrow \quad \swarrow{\scriptstyle \mu} & \\ & X & \end{array}$$

and

$$\begin{array}{ccc} B \otimes_A B \otimes_A X & \xrightarrow{m \otimes X} & B \otimes_A X \\ {\scriptstyle B \otimes \mu}\downarrow & & \downarrow{\scriptstyle \mu} \\ B \otimes_A X & \xrightarrow{\mu} & X \end{array}$$

are commutative.

As an A-module, the module $B \otimes_A X \cong X \oplus (M \otimes_A X)$. Therefore $\mu = (r,\alpha)$ where $r\colon X \longrightarrow X$ and $\alpha\colon M \otimes_A X \longrightarrow X$. The commutativity of the first diagram implies $r = id_X$. Then the commutativity of the second yields the relation $\alpha \cdot (M \otimes \alpha) = 0$. Thus, the B-module X determines uniquely the homomorphism $\alpha\colon M \otimes_A X \longrightarrow X$ with the relation $\alpha \cdot M \otimes \alpha = 0$, i.e. an element in $({}_{A}\underline{\underline{Mod}}) \ltimes (M \otimes_A -)$.

On the other hand, if we have an A-module X and such a morphism $\alpha\colon M \otimes_A X \longrightarrow X$, we can build $\mu\colon B \otimes_A X \longrightarrow X$ by defining $\mu = (id,\alpha)$. Then X obtains a left B-module structure (precisely: $(a,m) \cdot x = ax + \alpha(m \otimes x)$ for $a \in A$, $m \in M$ and $x \in X$).

We have partly demonstrated the claim: The categories ${}_{A \ltimes M}\underline{\underline{Mod}}$ and $({}_{A}\underline{\underline{Mod}}) \ltimes (M \otimes_A -)$ are isomorphic. The functors T, U, Z, and C have concrete realizations. Notice that the projection $B \xrightarrow{\pi} A$ is a ring homomorphism with M as kernel (M considered as an ideal in B). If X is an A-module, then ZX is X considered as a B-module, while $TX = B \otimes_A X$.

On the other hand, if X is a B-module, then UX is X considered as an A-module (through ι) while $CX = X/MX$. The functor $M \otimes_A - : {}_{A}\underline{\underline{Mod}} \longrightarrow {}_{A}\underline{\underline{Mod}}$ preserves arbitrary sums. This is not the case for all right exact functors. We state a result which implies that a right exact functor does preserve sums.

<u>Proposition</u> 1.13. <u>Suppose</u> $F\colon \underline{\underline{A}} \longrightarrow \underline{\underline{A}}$ <u>is a right exact functor</u>. <u>Then</u> $\underline{\underline{A}} \ltimes F$ <u>is the category of left modules over a ring if and only if</u> $\underline{\underline{A}}$ <u>is the category of modules over a ring and</u> F <u>is tensor product with a bimodule</u>.

The proof is an exercise in category theory. A very brief outline should suffice. Since $\underline{\underline{A}} \ltimes F$ has arbitrary sums and products and a small projective generator, so does $\underline{\underline{A}}$. Furthermore, the functor F will preserve arbitrary sums. But then $\underline{\underline{A}}$ is a category of modules, by Morita theory, and F is tensor product by Watt's theorem [71].

Palmer and Roos [57] have described flat left modules over the ring $A \ltimes M$, a description made possible by the equivalence: A left B-module X is flat if and only if $Hom_{\mathbb{Z}}(X, \mathbb{Q}/\mathbb{Z})$ is injective. Flat objects exist only in cases where a tensor product exists, thus a general description of a flat object in $\underline{\underline{A}} \times F$ cannot be made. This is the result of Palmer and Roos.

Proposition 1.14. Suppose α: $M \otimes_A X \longrightarrow X$ represents a left $A \ltimes M$-module. Then α is flat if and only if $C\alpha$ is flat and $M \otimes_A M \otimes_A X \xrightarrow{M \otimes \alpha} M \otimes_A X \longrightarrow X$ is exact.

Proof. Let $\underline{\underline{A}}$ denote the category ${}_A\underline{\underline{\mathrm{Mod}}}$ while $\underline{\underline{R}}$ denotes the category $\underline{\underline{\mathrm{Mod}}}_A$. Let $F = M \otimes_A -$ and $G = \mathrm{Hom}_A(M,-)$ on $\underline{\underline{R}}$. If α: $M \otimes_A X \longrightarrow X$, then

$$\mathrm{Hom}_{\mathbb{Z}}(\alpha, \mathbb{Q}/\mathbb{Z}): \ \mathrm{Hom}_{\mathbb{Z}}(X, \mathbb{Q}/\mathbb{Z}) \longrightarrow \mathrm{Hom}_{\mathbb{Z}}(M \otimes_A X, \mathbb{Q}/\mathbb{Z}).$$

But
$$\mathrm{Hom}_{\mathbb{Z}}(M \otimes_A X, \mathbb{Q}/\mathbb{Z}) \cong \mathrm{Hom}_A(M, \mathrm{Hom}_{\mathbb{Z}}(X, \mathbb{Q}/\mathbb{Z})).$$

Hence, $\mathrm{Hom}_{\mathbb{Z}}(\alpha, \mathbb{Q}/\mathbb{Z})$ becomes an object in $G \rtimes \underline{\underline{R}}$ (which is the category of right $A \ltimes M$-modules). Let $X^* = \mathrm{Hom}_{\mathbb{Z}}(X, \mathbb{Q}/\mathbb{Z})$ and $\alpha^* = \mathrm{Hom}_{\mathbb{Z}}(\alpha, \mathbb{Q}/\mathbb{Z})$.

Then α is flat if and only if α^* is injective. But α^* is injective if and only if $\ker \alpha^*$ is injective and the sequence $0 \longrightarrow \ker \alpha^* \longrightarrow X^* \xrightarrow{\alpha^*} GX^* \xrightarrow{G\alpha^*} G^2X^*$ is exact (Corollary 1.6 d)). But $\ker \alpha^* = (\mathrm{coker}\ \alpha)^*$. Also a sequence $X \longrightarrow Y \longrightarrow Z$ is exact if and only if $Z^* \longrightarrow Y^* \longrightarrow X^*$ is exact. Hence α^* is injective if and only if

$$M \otimes_A M \otimes_A X \xrightarrow{M \otimes \alpha} M \otimes_A X \xrightarrow{\alpha} X \longrightarrow \mathrm{coker}\ \alpha \longrightarrow 0$$

is exact and $\mathrm{coker}\ \alpha$ is flat. QED.

Since we will use this result also for the particular case of a triangular matrix ring $\begin{pmatrix} R & 0 \\ M & S \end{pmatrix}$, we will state the corresponding result.

Proposition 1.14 (bis.). Suppose α: $M \otimes_R X \longrightarrow Y$ represents a left module over $\begin{pmatrix} R & 0 \\ M & S \end{pmatrix}$. Then α is flat if and only if

a) X is flat as an R-module,

b) $\mathrm{coker}\ \alpha$ is flat as an S-module, and

c) α is an injection.

Proof. The homomorphism $\tilde{\alpha}$: $\tilde{F}(X,Y) \longrightarrow (X,Y)$ is given by

$(0,\alpha)$: $(0, M \otimes X) \longrightarrow (X,Y)$. Now $\tilde{\alpha}$ is flat if and only if $\tilde{F}^2(X,Y) \xrightarrow{\tilde{F}\tilde{\alpha}} \tilde{F}(X,Y) \xrightarrow{\tilde{\alpha}} (X,Y)$ is exact and coker $\tilde{\alpha}$ is flat. But coker $\tilde{\alpha} = (X, \text{coker } \alpha)$, while $\tilde{F}^2 = 0$. QED.

Let us use Proposition 1.11 in order to construct the injective envelope of the ring $A \ltimes M$. The ring $A \ltimes M$, as an object in $({}_A\underline{\underline{\text{Mod}}}) \ltimes (M \otimes_A -)$ is the extension of A, i.e. the object TA. (Denote $M \otimes_A -$ by F in order to facilitate this description.)

So

$$A \ltimes M = \begin{array}{c} F(A \oplus FA) \\ \Big\downarrow \begin{pmatrix} 0 & 0 \\ 1 & 0 \end{pmatrix} \\ A \oplus FA \end{array} .$$

In the category $\quad \text{Hom}_A(M,-) \rtimes {}_A\underline{\underline{\text{Mod}}}$,

this is the object

$$\begin{array}{c} A \oplus FA \\ \Big\downarrow \begin{pmatrix} 0 & 0 \\ \lambda & 0 \end{pmatrix} \\ GA \oplus GFA \end{array}$$

or

$$\begin{array}{c} A \oplus M \otimes_A A \\ \Big\downarrow \begin{pmatrix} 0 & 0 \\ \lambda & 0 \end{pmatrix} \\ \text{Hom}_A(M,A) \oplus \text{Hom}_A(M, M \otimes_A A). \end{array}$$

Now $\ker \lambda = \{a \in A: \ Ma = 0\}$ which is the right annihilator, rt. Ann M, of M.

According to Corollary 1.7b, the injective envelope of this is the object

$$H(E(\text{rt. Ann } M) \oplus M)).$$

This is the object

$$\begin{array}{c} \text{Hom}_A(M, E(\text{rt. Ann } M) \oplus E(M)) \oplus E(\text{rt. Ann } M) \oplus E(M) \\ \searrow = \\ \text{Hom}_A(M, \text{Hom}_A(M, E(\text{rt. Ann } M) \oplus E(M)) \oplus \text{Hom}_A(M, E(\text{rt. Ann } M) \oplus E(M)). \end{array}$$

(Let $I = E((\text{rt. Ann } M) \oplus M)$. This is the object

$$\begin{array}{c} GI \oplus I \\ \downarrow \begin{pmatrix} 0 & 0 \\ 1 & 0 \end{pmatrix} \\ G^2I \oplus GI \text{ .)} \end{array}$$

The homomorphism $A \oplus M \longrightarrow GI \oplus I$ is induced by $M \longrightarrow E(M)$, by rt. Ann $M \longrightarrow E(\text{rt. Ann } M)$ and by the composition

$$A \longrightarrow \text{Hom}_A(M,M) \longrightarrow \text{Hom}_A(M,E(M)).$$

Now return to $({}_A\underline{\underline{\text{Mod}}}) \ltimes (M \otimes_A -)$, to get the object

$$\begin{array}{c} FGI \oplus FI \\ \downarrow \begin{pmatrix} 0 & 0 \\ \rho & 0 \end{pmatrix} \\ GI \oplus I \end{array}$$

where ρ: $M \otimes \text{Hom}_A(M,X) \longrightarrow X$ is the usual trace map. This is

$$\begin{array}{c} (M \otimes_A \text{Hom}_A(M,E(\text{rt. Ann } M) \oplus E(M))) \oplus (M \otimes (E(\text{rt. Ann } M) \oplus E(M))) \\ \downarrow \begin{pmatrix} 0 & 0 \\ \rho & 0 \end{pmatrix} \\ (\text{Hom}_A(M,E(\text{rt. Ann } M) \oplus E(M))) \oplus (E(\text{rt. Ann } M) \oplus E(M)). \end{array}$$

As an abelian group, rather than a map, the injective envelope of $A \times M$ is then

$$\text{Hom}_A(M,E(\text{rt. Ann } M) \oplus E(M)) \oplus E(\text{rt. Ann } M) \oplus E(M).$$

The module action is given by

$$(a,m)\ (f,x) = (af, mf + ax)$$

for $\qquad a \in A,\ m \in M,\ f\colon\ M \longrightarrow I$ and $x \in I$.

The injection $A \ltimes M \longrightarrow GI \oplus I$ has been mentioned before.

A particular example, to which we will be returning in later sections, is that of a Dedekind domain A with field of quotients K. The functor $\mathrm{Ext}_A^1(M,-)$ is right exact for any M. Let $F = \mathrm{Ext}_A^1(K,-)$. Then F preserves finite, but not infinite sums. We then know that $({}_A\underline{\underline{\mathrm{Mod}}}) \ltimes F$ is not a category of modules. We invite the interested reader to consider the functor $T = \mathrm{Tor}_1^A(K/A,-)$ and the associated category

$$T \rtimes {}_A\underline{\underline{\mathrm{Mod}}} \, .$$

We will also study the category

$$\underline{\underline{\mathrm{Map}}}(F_A\underline{\underline{\mathrm{Mod}}}, {}_K\underline{\underline{\mathrm{Mod}}}) \, .$$

One further result.

<u>Proposition</u> 1.15. <u>The trivial extension</u> $A \ltimes M$ <u>is left perfect if and only if</u> A <u>is left perfect</u>.

Proof. Corollary 1.7a). QED.

Section 2. Coherence

A ring R is said to be left coherent if every finitely generated left ideal is finitely presented. This condition is equivalent to the two equivalent conditions: Each finitely generated submodule of a finitely presented left R-module is finitely presented. Each product of flat right R-modules is flat.

The concept of coherence can be generalized.

Suppose $\underline{\underline{P}}$ is a small additive category. We denote by $\underline{\underline{Ab}}$ the category of abelian groups (although a suitable category of modules could be taken for $\underline{\underline{Ab}}$). By $[\underline{\underline{P}},\underline{\underline{Ab}}]$ we denote the category of additive covariant functors from $\underline{\underline{P}}$ to $\underline{\underline{Ab}}$ and by $[\underline{\underline{P}}^{op},\underline{\underline{Ab}}]$ we denote the category of additive contravariant functors. For each P in $\underline{\underline{P}}$ we have the representable functors $\underline{\underline{P}}(P,-) = h^P$ in $[\underline{\underline{P}},\underline{\underline{Ab}}]$ and $\underline{\underline{P}}(-,P) = h_P$ in $[\underline{\underline{P}}^{op},\underline{\underline{Ab}}]$. By Yoneda's lemma, the functors $h\colon \underline{\underline{P}}^{op} \longrightarrow [\underline{\underline{P}},\underline{\underline{Ab}}]$ and $h\colon \underline{\underline{P}} \longrightarrow [P^{op},\underline{\underline{Ab}}]$ identify $\underline{\underline{P}}^{op}$ (resp.: $\underline{\underline{P}}$) as a full additive subcategory of $[\underline{\underline{P}},\underline{\underline{Ab}}]$ (resp.: $[\underline{\underline{P}}^{op},\underline{\underline{Ab}}]$) consisting of projective objects.

Suppose $\underline{\underline{A}}$ is an abelian category which has a small full additive subcategory $\underline{\underline{P}}$ whose objects are projective in $\underline{\underline{A}}$.

Definition. Let A be an object in $\underline{\underline{A}}$.

a) The object A is of finite $\underline{\underline{P}}$-type if there is an epimorphism $P \longrightarrow A$ for some P in $\underline{\underline{P}}$.

b) The object is of finite $\underline{\underline{P}}$-presentation if there is an exact sequence $Q \longrightarrow P \longrightarrow A \longrightarrow 0$ for P and Q in $\underline{\underline{P}}$.

c) The object is $\underline{\underline{P}}$-coherent if it is of finite P-type and every subobject of finite $\underline{\underline{P}}$-type is of finite $\underline{\underline{P}}$-presentation.

d) The object is pseudo $\underline{\underline{P}}$-coherent if every subobject of finite $\underline{\underline{P}}$-type is coherent.

Let $\underline{\underline{\mathrm{Coh}}}_{\underline{\underline{P}}}\underline{\underline{A}}$ denote the full additive subcategory of $\underline{\underline{P}}$-coherent objects. It is abelian.

We have a functor $h\colon \underline{\underline{A}} \longrightarrow [\underline{\underline{P}}^{op},\underline{\underline{Ab}}]$ given by $h_A = \underline{\underline{A}}(-,A)$ for each object (resp: morphism) A in $\underline{\underline{A}}$. Since the objects in $\underline{\underline{P}}$ are projective, the functor h is exact.

Again, by Yoneda's lemma, the functor h identifies the objects of finite $\underline{\underline{P}}$-presentation in $\underline{\underline{A}}$ with the objects of finite $\underline{\underline{P}}$-presentation in $[\underline{\underline{P}}^{op},\underline{\underline{Ab}}]$. For suppose $P \longrightarrow Q \longrightarrow A \longrightarrow 0$ is exact in A. Then $h_P \longrightarrow h_Q \longrightarrow h_A \longrightarrow 0$ is exact in $[\underline{\underline{P}}^{op},\underline{\underline{Ab}}]$, so h_A is of finite $\underline{\underline{P}}$-presentation. Conversely, let P and Q be objects in $\underline{\underline{P}}$. Since $[h_P,h_Q] = \underline{\underline{A}}(P,Q)$, the cokernel of a morphism $h_P \longrightarrow h_Q$ is representable by the cokernel of the corresponding $P \longrightarrow Q$ in $\underline{\underline{A}}$. Furthermore the object A is $\underline{\underline{P}}$-coherent if and only if h_A is $\underline{\underline{P}}$-coherent.

Thus the categories $\underline{\underline{\mathrm{Coh}}}_{\underline{\underline{P}}}\underline{\underline{A}}$ and $\underline{\underline{\mathrm{Coh}}}_{P}[\underline{\underline{P}}^{op},\underline{\underline{Ab}}]$ are equivalent.

<u>Definition</u>. We say that $\underline{\underline{P}}$ is <u>left coherent</u> if $\underline{\underline{P}} \subset \underline{\underline{\mathrm{Coh}}}_{P}[\underline{\underline{P}}^{op},\underline{\underline{Ab}}]$. Correspondingly we say $\underline{\underline{P}}$ is <u>right coherent</u> if $\underline{\underline{P}}^{op} \subset \underline{\underline{\mathrm{Coh}}}_{P^{op}}[\underline{\underline{P}},\underline{\underline{Ab}}]$, and that $\underline{\underline{P}}$ is <u>coherent</u> if $\underline{\underline{P}}$ is both left and right coherent.

There is a tensor product

$$\otimes\colon\ [\underline{\underline{P}},\underline{\underline{Ab}}] \times [\underline{\underline{P}}^{op},\underline{\underline{Ab}}] \longrightarrow \underline{\underline{Ab}}$$

which represents the functor

$$(R,L)_G \longmapsto [L,\mathrm{Hom}_{\mathbb{Z}}(R,G)]$$

for any abelian group G. That is, for any abelian group G, the abelian group of natural transformations

$$[L,\mathrm{Hom}_{\mathbb{Z}}(R-,G)] \cong \mathrm{Hom}_{\mathbb{Z}}(R \otimes_{\underline{\underline{P}}} L,G).$$

(See, for example Oberst and Rohrl [53] or Mitchell [44].)

The tensor product has derived functors $\mathrm{Tor}_I^{\underline{\underline{P}}}(\ ,\ )$ and in terms of these derived functors one can define flatness. The crucial result, relating the categorical abstractions to the more concrete ring theory is the following result due to Oberst and Rohrl [53].

Theorem 2.1. *The following conditions on* $\underline{\underline{P}}$ *are equivalent.*

a) *Each* h_P *is coherent.*

b) *Each finitely* $\underline{\underline{P}}$*-presented contravariant functor is coherent.*

c) *The product of flat covariant functors is flat.*

We thus have an intrinsic characterization of left coherent small additive categories, and thus a condition in order that $\underline{\underline{P}} \subset \underline{\underline{\mathrm{Coh}}}_{\underline{\underline{P}}}(\underline{\underline{A}})$.

However, we do not use this in our next result, preferring to work inside $\underline{\underline{A}}$ rather than in the functor categories.

Theorem 2.2. *Suppose* $\underline{\underline{P}}$ *is a small additive full subcategory of projectives in* $\underline{\underline{A}}$. *Suppose* $F: \underline{\underline{A}} \longrightarrow \underline{\underline{A}}$ *is right exact. Let* $T(\underline{\underline{P}})$ *denote the small additive full subcategory of extensions of* $\underline{\underline{P}}$ *to* $\underline{\underline{A}} \ltimes F$. *Then* $T(\underline{\underline{P}})$ *is coherent in* $\underline{\underline{A}} \ltimes F$ *if and only if the following conditions are satisfied:*

a) *The category* $\underline{\underline{P}}$ *is coherent in* $\underline{\underline{A}}$.

For all $\underline{\underline{P}}$*-coherent objects* A *in* $\underline{\underline{A}}$,

b) *the objects* $L_iF(A)$ *are* $\underline{\underline{P}}$*-coherent for all* $i > 0$ *and*

c) *if* B *is a subobject of* FA *of finite* $\underline{\underline{P}}$*-type, then* B *is coherent and* FB *is of finite* $\underline{\underline{P}}$*-type.*

Before starting the proof we note that an object α in $\underline{\underline{A}} \ltimes F$ is of finite $T(\underline{\underline{P}})$-type if and only if $C\alpha$ is of finite $\underline{\underline{P}}$-type.

Proof. Suppose $T(\underline{\underline{P}})$ is coherent. We first show that $\underline{\underline{P}}$ is coherent. Let $\alpha: P \longrightarrow Q$ be a morphism. Then $\ker\begin{pmatrix}\alpha & 0\\ 0 & F\alpha\end{pmatrix}: TP \longrightarrow TQ$ is of finite $T(\underline{\underline{P}})$-type. But $\ker\begin{pmatrix}\alpha & 0\\ 0 & F\alpha\end{pmatrix} = \begin{pmatrix}0 & 0\\ 1 & 0\end{pmatrix}: F(\ker \alpha) \oplus F(\ker F\alpha) \longrightarrow \ker \alpha \oplus \ker F\alpha$ and hence $\ker \alpha$ is a direct summand of $C(\ker\begin{pmatrix}\alpha & 0\\ 0 & F\alpha\end{pmatrix})$ and is therefore of finite $\underline{\underline{P}}$-type.

Now suppose A is coherent. Then there is a projective resolution $P_\cdot \longrightarrow A$ of A by objects in $\underline{\underline{P}}$. Then the complex $T(P_\cdot)$ consists of coherent objects in $\underline{\underline{A}} \ltimes F$. Thus its homology modules are coherent in $\underline{\underline{A}} \ltimes F$. But its homology modules are just $Z(L_iF(A))$ for $i > 0$ by Proposition 1.9. Hence, each $L_iF(A)$ is coherent for $i > 0$.

Suppose $x: P \longrightarrow FA$. Then we get a morphism $\begin{pmatrix} 0 & 0 \\ x & 0 \end{pmatrix}: TP \longrightarrow TA$. Since A is coherent and T is right exact, the object TA is coherent in $\underline{\underline{A}} \ltimes F$. Since $0 \longrightarrow \ker x \oplus FP \longrightarrow TP \longrightarrow TA$ is exact, the object

$$F(\ker x \oplus FP) \xrightarrow{\begin{pmatrix} 0 & 0 \\ 1 & 0 \end{pmatrix}} \ker x \oplus FP$$

is coherent in $\underline{\underline{A}} \ltimes F$. Hence, the cokernel, which is $\ker x \oplus F(P/\ker x)$, is of finite $\underline{\underline{P}}$-type.

Thus we have verified the conditions a), b) and c).

Suppose now that a), b) and c) hold. We want to show that $T(\underline{\underline{P}})$ is coherent. We must show that the kernel of a morphism $T(P) \longrightarrow T(Q)$ is of finite $T(\underline{\underline{P}})$-type. Such a morphism is given by a matrix $\mu = \begin{pmatrix} \alpha & 0 \\ x & F\alpha \end{pmatrix}: P \oplus FP \longrightarrow Q \oplus FQ$. Let Z denote the object $\ker \mu$ in $\underline{\underline{A}}$. There is a unique $h: FZ \longrightarrow Z$ such that the object h in $\underline{\underline{A}} \ltimes F$ is the kernel of μ. To show that h is of finite $T(\underline{\underline{P}})$-type, it is sufficient to show that $C(h)$ is of finite $\underline{\underline{P}}$-type in $\underline{\underline{A}}$. Denote $C(h)$ by C.

Let $\begin{pmatrix} u \\ v \end{pmatrix} : Z \longrightarrow P \oplus FP$ be the kernel. Then we know that

$$\begin{pmatrix} u \\ v \end{pmatrix} \cdot h = \begin{pmatrix} 0 & 0 \\ 1 & 0 \end{pmatrix} \cdot \begin{pmatrix} Fu \\ Fv \end{pmatrix} .$$

Hence $uh = 0$ and $vh = Fu$. The first relation tells us that u factors uniquely through C and that h factors uniquely through $\ker u$. We also know that $\begin{pmatrix} \alpha & 0 \\ x & F\alpha \end{pmatrix} \begin{pmatrix} u \\ v \end{pmatrix} = \begin{pmatrix} 0 \\ 0 \end{pmatrix}$ and hence that $\alpha u = 0$ and $x \cdot u + F\alpha \cdot v = 0$. Thus u factors uniquely through $\ker \alpha \longrightarrow P$ and Z is then the pullback of the pair $(x|_{\ker \alpha}, F\alpha)$. Let x' denote the morphism $x|_{\ker \alpha}$, let X denote $\ker x'$ and Y denote cokernel x'. The other notation in the commutative diagram below is obvious.

Consider the diagram

$$\begin{array}{ccc} 0 & & 0\\ \downarrow & & \downarrow \\ X & = & X \\ \downarrow & & \downarrow \\ Z & \xrightarrow{-v} & FP \\ {\scriptstyle u}\downarrow & & \downarrow{\scriptstyle F\alpha} \\ \ker\alpha & \xrightarrow{x'} & FQ \\ \downarrow & & \downarrow \\ Y & \longrightarrow & FB \\ \downarrow & & \downarrow \\ 0 & & 0 \; . \end{array}$$

Since Z is the pullback, the kernel of $F\alpha$ is identified with the kernel of u while the cokernel Y of u is mapped monomorphically to $\operatorname{coker} F\alpha$, which is $F(B)$ with $B = \operatorname{coker} \alpha$. Also let $\operatorname{im} \alpha$ be denoted by A. We get the commutative diagram

$$\begin{array}{ccccccccccc} 0 & \longrightarrow & X & \longrightarrow & FP & \xrightarrow{F\alpha} & FQ & \longrightarrow & FB & \longrightarrow & 0 \\ & & & & \downarrow & & \downarrow{\scriptstyle =} & & \downarrow{\scriptstyle =} & & \\ 0 & \longrightarrow & L_1FB & \longrightarrow & FA & \longrightarrow & FQ & \longrightarrow & FB & \longrightarrow & 0 \end{array}$$

and a uniquely induced epimorphism $X \longrightarrow L_1FB$.

By the previous remarks we get also the diagram (commutative as always)

$$\begin{array}{ccccccccccc} 0 & \longrightarrow & X & \longrightarrow & Z & \xrightarrow{u} & \ker\alpha & \longrightarrow & Y & \longrightarrow & 0 \\ & & \downarrow & & \downarrow & & \downarrow{\scriptstyle =} & & \downarrow{\scriptstyle =} & & \\ 0 & \longrightarrow & U & \longrightarrow & C & \longrightarrow & \ker\alpha & \longrightarrow & Y & \longrightarrow & 0 \end{array}$$

and then a uniquely induced epimorphism $X \longrightarrow U$. There is then the associated diagram

$$\begin{array}{ccccccccccc} 0 & \longrightarrow & U & \longrightarrow & C & \longrightarrow & \ker\alpha & \longrightarrow & Y & \longrightarrow & 0 \\ & & \downarrow & & \downarrow & & \downarrow{\scriptstyle x'} & & \downarrow & & \\ 0 & \longrightarrow & L_1FB & \longrightarrow & FA & \longrightarrow & FQ & \longrightarrow & FB & \longrightarrow & 0 \end{array}$$

with induced morphisms $C \longrightarrow FA$ and $U \longrightarrow L_1FB$.

Since $vh = Fu$ we get the diagram

$$\begin{array}{ccc} FZ & \xrightarrow{-Fu} & F(\ker \alpha) \\ \downarrow{\scriptstyle h} & & \downarrow{\scriptstyle h'} \\ Z & \xrightarrow{-v} & FP \end{array}$$

where both h and h' factor through X. Hence, we have induced the diagram

$$\begin{array}{ccccccc} FZ & \xrightarrow{-Fu} & F(\ker \alpha) & \longrightarrow & FY & \longrightarrow & 0 \\ \downarrow & & \downarrow & & & & \\ X & \xrightarrow{=} & X & & & & \\ \downarrow & & \downarrow & & & & \\ U & \longrightarrow & L_1FB & & & & \\ \downarrow & & \downarrow & & & & \\ 0 & & 0 & & & & \end{array}$$

with exact rows and columns. There is thus induced a morphism $FY \longrightarrow U$ such that $FY \longrightarrow U \longrightarrow L_1FB \longrightarrow 0$ is exact.

Putting all of this arrow theory together gives the two exact sequences

$$\begin{array}{ccccccccccc} & & FY & & & & & & & & \\ & & \downarrow & & & & & & & & \\ 0 & \longrightarrow & U & \longrightarrow & C & \longrightarrow & \ker \alpha & \longrightarrow & Y & \longrightarrow & 0 \\ & & \downarrow & & & & & & & & \\ & & L_1FB & & & & & & & & \\ & & \downarrow & & & & & & & & \\ & & 0. & & & & & & & & \end{array}$$

We now apply the hypotheses. Since $\underline{\underline{P}}$ is coherent, the objects $\ker \alpha$, A and B are coherent. Thus $\ker \alpha$ is of finite $\underline{\underline{P}}$-type, which implies that Y is of finite $\underline{\underline{P}}$-type. But Y is a subobject of FB and hence coherent. Therefore $\ker(\ker \alpha \longrightarrow Y)$ is at least of finite $\underline{\underline{P}}$-type. Also FY is of finite $\underline{\underline{P}}$-type. Since L_1FB is coherent, it is of finite $\underline{\underline{P}}$-type. Hence U is squeezed between two objects of finite $\underline{\underline{P}}$-type and therefore is itself of finite $\underline{\underline{P}}$-type. But then C is squeezed between U and the kernel of $\ker \alpha \longrightarrow Y$ and hence C is also of finite $\underline{\underline{P}}$-type. QED.

Remark. It appears that the coherence of the derived objects L_1FB is not used in the proof. However, each L_1FB is a subobject of an FP for P in $\underline{\underline{P}}$. Thus, together with condition c) one gets that L_1FB is $\underline{\underline{P}}$-coherent if and only if it is of finite $\underline{\underline{P}}$-type.

We include a complete diagram which perhaps makes clearer the argument used above.

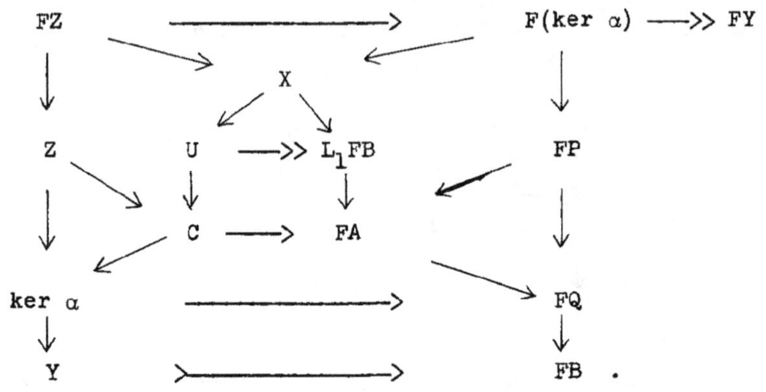

In case $\underline{\underline{A}} = {}_A\underline{\underline{Mod}}$ and $F = M \otimes_A -$, we get conditions which are necessary and sufficient in order that the ring $A \ltimes M$ be left coherent. In particular the result of Roos [65] on the coherence of $\begin{pmatrix} R & N \\ M & S \end{pmatrix}_{(0,0)}$ is a corollary.

Corollary 2.3. The ring $\begin{pmatrix} R & N \\ M & S \end{pmatrix}_{(0,0)}$ *is left coherent if and only if:*

a) *The rings* R *and* S *are left coherent.*

And for each finitely presented left R-*module* A *and finitely presented left* S-*module* B,

b) *the left modules* $\mathrm{Tor}_i^R({}_SM,A)$ *and* $\mathrm{Tor}_i^S({}_RN,B)$ *are finitely presented for all* $i > 0$, *and*

c) *if* C *is a finitely generated left sub* S-*module of* $M \otimes_R A$ (resp.: sub R-module of $N \otimes_S B$) *then* C *is finitely presented and* $N \otimes_S C$ (resp.: $M \otimes_R C$) *is finitely generated.*

Proof. In this case $\underline{\underline{A}} = {}_{R \times S}\underline{\underline{Mod}}$ while $F = (M \oplus N) \otimes -$. However each $R \times S$-module is a module of the form $A \times B$, where A is an R-module and B is an S-module. Then $F(A \times B) = (N \otimes_S B) \times (M \otimes_R A)$ and $L_iF(A \times B) \cong \mathrm{Tor}_i^S(N,B) \times \mathrm{Tor}_i^R(M,A)$ for $i > 0$.

Therefore the corollary follows. QED.

As an example of an application of this result, consider the field k and the coherent k-algebra $k[\{X_i\}_{i \in S}]$, which we denote by A. (Of course we assume S is infinite). Let k be considered as an A-module by the usual augmentation $A \longrightarrow k$ obtained by sending each X_i to 0. Then the ring $\begin{pmatrix} A & 0 \\ k & k \end{pmatrix}$ is left coherent but not right coherent.

As another example, suppose A is a commutative noetherian ring. Let E be an indecomposable injective A-module. Then E is the injective envelope of $A/\underline{p}$ for some prime ideal $\underline{p}$ in Spec A. Furthermore $\{\underline{p}\} = \text{Ass } E$. When does the functor $E \otimes_A -$ satisfy the conditions b) and c) of the theorem? The following discussion is to answer this question.

Since A is noetherian, condition b) is satisfied provided $\text{Tor}_i^A(E,M)$ is finitely generated for all $i > 0$ and all finitely generated A-modules M. Since E is injective, the module $\text{Tor}_i^A(E,M) \cong \text{Hom}_A(\text{Ext}_A^i(M,A),E)$ for all $i > 0$ [Cartan and Eilenberg, 14]. But E is already an $A_{\underline{p}}$-module. Hence $\text{Hom}_A(\text{Ext}_A^i(M,A),E) \cong \text{Hom}_{A_{\underline{p}}}(\text{Ext}_A^i(M,A)_{\underline{p}},E)$. Since $\text{Ext}_A^i(M,A)_{\underline{p}}$ is finitely generated, its E dual, which is $A_{\underline{p}} \otimes \text{Tor}_i^A(M,E)$, has the descending chain condition. Thus it is of finite type if and only if it is of finite length. But then $\text{Ext}_A^i(M,A)_{\underline{p}}$ is of finite length.

Lemma 2.4. The local noetherian ring B has $\text{Ext}_B^i(M,B)$ of finite length for all $i > 0$ and all finitely generated B-modules M if and only if $B_{\underline{q}}$ is self injective for all $\underline{q}$ in Spec $B-\{\underline{m}\}$. In particular $\dim B \leq 1$ when one of these conditions is satisfied.

Proof. Suppose $\underline{q} \in \text{Spec } B-\{\underline{m}\}$. Then

$$\text{Ext}^i_{B_{\underline{q}}}(k(\underline{q}),B_{\underline{q}}) \cong \text{Ext}_B^i(B/\underline{q},B)_{\underline{q}} = 0$$

for all $i > 0$. But then $B_{\underline{q}}$ is self injective. QED.

Corollary 2.5. If E is the injective envelope of the A-module $A/\underline{p}$ and if $\text{Tor}_i^A(E,M)$ is of finite type for all $i > 0$ and all finitely generated A-modules, then $A_{\underline{q}}$ is self injective for all $\underline{q} \subsetneq \underline{p}$

and $\underline{p}$ is a maximal ideal, or ht $p = 0$ and $A_{\underline{p}}$ is Gorenstein. The converse holds as well.

Proof. The first part of the conclusion follows from the lemma. As for the second, if $\mathrm{Ext}^i_{A_{\underline{p}}}(k(\underline{p}),A_{\underline{p}}) = 0$ for some $i > 0$, then $\mathrm{Tor}_i^A(A/\underline{p},E) = 0$. It is of finite length over $A_{\underline{p}}$ and so admits $k(\underline{p})$ as a submodule. Hence $k(\underline{p})$ is finitely generated as an A-module. Therefore $A/\underline{p} \cong k(\underline{p})$ and so $\underline{p}$ is a maximal ideal. Otherwise $\mathrm{Ext}^i_{A_{\underline{p}}}(k(\underline{p}),A_{\underline{p}}) = 0$ for all $i > 0$ and so $A_{\underline{p}}$ is self-injective. The converse is easily verified. QED.

Now consider condition c). It is enough to look at finitely generated submodules of E. If X is a finitely generated submodule of E, then $X \subseteq X_{\underline{p}} \subseteq E$ and $X_{\underline{p}}$ has finite length as an $A_{\underline{p}}$-module. Now $E \otimes_A X \cong E \otimes_A X_{\underline{p}}$ as A-modules. Furthermore $E \otimes_A X_{\underline{p}} \cong \mathrm{Hom}_{A_{\underline{p}}}(\mathrm{Hom}_{A_{\underline{p}}}(X_{\underline{p}},A_{\underline{p}}),E)$. So $E \otimes_A X_{\underline{p}} \neq 0$ if and only if depth $A_{\underline{p}} = 0$.

If depth $A_{\underline{p}} = 0$, then $E \otimes_A k(\underline{p}) \neq 0$ and is of finite type as an A-module. Once again we conclude that $\underline{p}$ is a maximal ideal. The conclusions then are summed up in the next statement.

Proposition 2.6. Suppose E is the injective envelope of $A/\underline{p}$. Then E satisfies the conditions

b) $\mathrm{Tor}_i^A(E,M)$ is of finite type for all finitely generated M

and c) if $X \subseteq E \otimes M$ and is of finite type then $E \otimes_A X$ is of finite type if and only if

i) The height ht $p \leq 1$ and $\underline{p}$ is a maximal prime ideal.

ii) If $\underline{q} \subsetneq \underline{p}$, then $A_{\underline{q}}$ is Gorenstein. QED.

For a more general injective module we can first decompose it into a direct sum of indecomposable injective modules. Each $E(A/\underline{p})$ can occur but a finite number of times, and must satisfy the conditions of the proposition. The details are left for the reader.

As another example, suppose A is a Dedekind domain with field of quotients K. Let $\underline{\underline{A}}$ and $\underline{\underline{K}}$ denote the categories of A-modules and K-modules respectively. Let $\mathrm{Ext}^1_A(K,-)\colon \underline{\underline{A}} \longrightarrow \underline{\underline{K}}$ be the functor and consider the category which we denote by $\underline{\underline{\Lambda}}$, that is

$$\underline{\underline{\Lambda}} = \begin{pmatrix} \underline{\underline{A}} & 0 \\ \mathrm{Ext}^1_A(K,-) & \underline{\underline{K}} \end{pmatrix} = \underline{\underline{\mathrm{Map}}}(\mathrm{Ext}^1_A(K,-)\ \underline{\underline{\mathrm{Mod}}}_A,\ \underline{\underline{\mathrm{Mod}}}_K).$$

Denote by E the object $\mathrm{Ext}^1_A(K,A)$. Then $\Lambda = \begin{pmatrix} A & 0 \\ E & K \end{pmatrix}$ is a projective generator of $\underline{\underline{\Lambda}}$. Note that $\underline{\underline{\Lambda}}$ does not have arbitrary sums. It is not difficult to demonstrate the following statements.

Proposition 2.7. The ring Λ is left and right coherent. The category of coherent left Λ-modules is isomorphic to the category of coherent objects in $[\underline{\underline{P}}^{op},Ab]$ where $\underline{\underline{P}}$ is the additive subcategory of $\underline{\underline{\Lambda}}$ consisting of finite direct sums of copies of Λ. The category of coherent right Λ-modules is isomorphic to the category of coherent objects in $[\underline{\underline{P}},\underline{\underline{Ab}}]$. QED.

Perhaps it is wise to interpret Theorem 2.2 in terms of the category $\underline{\underline{\mathrm{Map}}}(F\underline{\underline{A}},\underline{\underline{B}})$. This is the category $(\underline{\underline{A}} \times \underline{\underline{B}}) \ltimes \tilde{F}$.

Suppose $\underline{\underline{P}}$ is a small additive full subcategory of projectives in $\underline{\underline{A}}$ and $\underline{\underline{Q}}$ the same in $\underline{\underline{B}}$. Then the small additive full subcategory $\underline{\underline{P}} \times \underline{\underline{Q}}$ in $\underline{\underline{A}} \times \underline{\underline{B}}$ is coherent if and only if $\underline{\underline{P}}$ is coherent in $\underline{\underline{A}}$ and $\underline{\underline{Q}}$ is coherent in $\underline{\underline{B}}$.

Now $L_i\tilde{F}(A,B) = (0,L_iFA)$ for all A in $\underline{\underline{A}}$ and B in $\underline{\underline{B}}$. And $\tilde{F}^2 = 0$. Thus, we see that $T(\underline{\underline{P}} \times \underline{\underline{Q}})$ is coherent if and only if $\underline{\underline{P}}$ and $\underline{\underline{Q}}$ are coherent and for all finitely $\underline{\underline{P}}$-presented objects A in $\underline{\underline{A}}$ the objects L_iFA are finitely $\underline{\underline{Q}}$-presented for $i > 0$ and FA is pseudo-coherent (every subobject of finite $\underline{\underline{Q}}$-type is of finite $\underline{\underline{Q}}$-presentation).

For a final example consider the von Neumann regular ring R. Now R is von Neumann regular if and only if every left (right) R-module is flat. Hence R is certainly coherent. Moreover a left R-module is of finite presentation if and only if it is projective of finite type. The R-bimodule M induces a functor $F = M \otimes_R -$ satisfying the conditions of Theorem 2.2 if and only if M is projective and of finite type as a left R-module. Thus $R \ltimes M$ is left coherent if and only if M is left coherent. We can also compute the flat dimension of $R \ltimes M$. In fact

$$\text{left flat dim } R \ltimes M = \inf \{n-1 : \underbrace{M \otimes M \otimes \ldots \otimes M}_{n} = 0\}.$$

We note that R is a Gorenstein ring. Hence we have a large class of non-noetherian Gorenstein rings.

Section 3. Duality and the Gorenstein property

Recall that $\underline{\underline{P}}$, the small additive category, is coherent (say left coherent) if each representable functor $h_P \in [\underline{\underline{P}}^{op},Ab]$ is a coherent object. Say that $\underline{\underline{P}}$ is right coherent if $\underline{\underline{P}}^{op}$ is coherent and that $\underline{\underline{P}}$ is coherent if $\underline{\underline{P}}$ is both left and right coherent.

If $\underline{\underline{P}}$ is coherent there are contravariant functors

$$\alpha: \underline{\underline{Coh}}[\underline{\underline{P}}^{op},\underline{\underline{Ab}}] \longrightarrow \underline{\underline{Coh}}[\underline{\underline{P}},\underline{\underline{Ab}}]$$

and

$$\beta: \underline{\underline{Coh}}[\underline{\underline{P}},\underline{\underline{Ab}}] \longrightarrow \underline{\underline{Coh}}[\underline{\underline{P}}^{op},\underline{\underline{Ab}}]$$

defined as follows. If $P \in \underline{\underline{P}}$, then $\alpha h_P = h^P$ while $\alpha h_f = h^f$ for a morphism f in $\underline{\underline{P}}$. Then α is extended to the finitely presented objects by insisting that it be left exact. So if $h_P \xrightarrow{h_f} h_Q \longrightarrow F \longrightarrow 0$ is exact in $[\underline{\underline{P}}^{op},\underline{\underline{Ab}}]$, then $\alpha F = \ker \alpha h_f$. The functor β has for values $\beta h^P = h_P$ and $\beta h^f = h_f$ and is also forced to be left exact.

The prototypes for α and β are obtained by considering the categories of left and right modules for the coherent ring R. Thus $\alpha = Hom_R(-,.R)$ acts on (coherent or finitely presented) left R-modules.

Returning to the general case, we see that α and β are contravariant functors adjoint on the right. That is, for each pair F,G of finitely presented functors, there is an isomorphism, natural in F and G,

$$Hom(G,\alpha F) \cong Hom(F,\beta G).$$

For example, we get

$$\mathrm{Hom}(h^P, \alpha h_Q) \cong \mathrm{Hom}(h^P, h^Q) \cong \underline{\underline{P}}(Q,P)$$

while

$$\mathrm{Hom}(h_Q, \beta h^P) \cong \mathrm{Hom}(h_Q, h_P) \cong \underline{\underline{P}}(Q,P).$$

For the prototype, we have

$$\varphi\colon\ \mathrm{Hom}_R(M.,\mathrm{Hom}_R(.N,.R)) \cong \mathrm{Hom}_R(.N,\mathrm{Hom}_R(M.,R.))$$

for finitely presented right (resp.: left) coherent modules $M.$ and $.N$. If we write homomorphisms on the left of elements in a right module and homomorphisms on the right of elements in a left module then we can describe φ acting on f by $(nf^{\varphi})(m) = n(f(m))$.

The functors α and β restrict to a perfect duality on the subcategories $\underline{\underline{P}}$ and $\underline{\underline{P}}^{op}$, respectively.

In general, if $\underline{\underline{A}}$ and $\underline{\underline{B}}$ are abelian categories and there are given contravariant, adjoint on the right, functors $\alpha'\colon \underline{\underline{A}} \longrightarrow \underline{\underline{B}}$ and $\beta'\colon \underline{\underline{B}} \longrightarrow \underline{\underline{A}}$, then the pair (α',β') is said to be a <u>pseudoduality</u> if there is a small additive coherent category $\underline{\underline{P}}$ and equivalences of categories

$$e\colon\ \underline{\underline{A}} \xrightarrow{\cong} \underline{\underline{\mathrm{Coh}}}[\underline{\underline{P}}^{op},\underline{\underline{Ab}}]$$

and

$$f\colon\ \underline{\underline{B}} \xrightarrow{\cong} \underline{\underline{\mathrm{Coh}}}[\underline{\underline{P}},\underline{\underline{Ab}}]$$

such that the diagrams

$$\begin{array}{ccc} \underline{\underline{A}} & \xrightarrow{\alpha} & \underline{\underline{B}} \\ e\downarrow & & \downarrow f \\ \underline{\underline{\mathrm{Coh}}}[\underline{\underline{P}}^{op},\underline{\underline{Ab}}] & \xrightarrow{\alpha} & \underline{\underline{\mathrm{Coh}}}[\underline{\underline{P}},\underline{\underline{Ab}}] \end{array} \quad \text{and} \quad \begin{array}{ccc} \underline{\underline{A}} & \xleftarrow{\beta'} & \underline{\underline{B}} \\ e\downarrow & & \downarrow f \\ \underline{\underline{\mathrm{Coh}}}[\underline{\underline{P}}^{op},\underline{\underline{Ab}}] & \xleftarrow{\beta} & \underline{\underline{\mathrm{Coh}}}[\underline{\underline{P}},\underline{\underline{Ab}}] \end{array}$$

are commutative. The terminology is mostly for convenience, in order to avoid the repeated use of the functor category notation.

We denote by $\underline{\underline{P}}$ the full small additive subcategory of projectives in $\underline{\underline{A}}$ and by $\underline{\underline{Q}}$ its dual in $\underline{\underline{B}}$. Then $\underline{\underline{A}} = \underline{\underline{\mathrm{Coh}}}_{\underline{\underline{P}}}\underline{\underline{A}}$ and $\underline{\underline{B}} = \underline{\underline{\mathrm{Coh}}}_{\underline{\underline{Q}}}\underline{\underline{B}}$.

Since α and β are adjoint, they are both left exact. Since each object in $\underline{\underline{A}}$ is finitely $\underline{\underline{P}}$-presented, we can take projective resolutions and then get the right derived functors $R^i\alpha$(resp.: $R^i\beta$). On the prototype, $R^i\alpha = \mathrm{Ext}^i_R(-,R.)$ while $R^i\beta = \mathrm{Ext}^i_R(-,.R)$. The reader interested only in ring theory is invited to interpret the remainder of this chapter in these terms.

(The case for $\aleph_\alpha$-coherent rings is more difficult and shows the necessity of the general theory. Suppose R is left $\aleph_\alpha$-coherent; that is each left ideal generated by $\aleph_\alpha$ elements is $\aleph_\alpha$ related. Let $\underline{\underline{P}}$ be the category of projective left R-modules with $\aleph_\alpha$ generators. There is a free R-module P such that each of these projective modules is a direct summand of P. Thus, each $\underline{\underline{P}}$-coherent left R-module A is of the form P/Px for some $x\colon P \longrightarrow P$. Let $E = \mathrm{End}_R P$. Then the pseudo dual of A is the kernel of the morphism $x_*\colon \mathrm{Hom}_R(P,-) \longrightarrow \mathrm{Hom}_R(P,-)$. There is first the problem to determine whether this is a coherent object. It is coherent if and only if there is an epimorphism $\mathrm{Hom}_R(P,-) \longrightarrow \mathrm{Hom}_R(A,-)$. Such an epimorphism is given by a morphism $a\colon A \longrightarrow P$. Equivalenty we should find a $g\colon P \longrightarrow P$ such that the sequence $E \xrightarrow{g^*} E \xrightarrow{f} E$ is exact. The study of these strange rings is left for another time. This leaves open the question whether there are applications of the pseudo duality other than to left and right coherent rings.)

For each presentation $P_1 \longrightarrow P_o \longrightarrow A \longrightarrow 0$ of an object A in $\underline{\underline{A}}$ we can associate the transpose object TA in $\underline{\underline{B}}$ defined to be the cokernel of $\alpha P_o \longrightarrow \alpha P_1$. Thus we have the exact sequence

$$0 \longrightarrow \alpha A \longrightarrow \alpha P_o \longrightarrow \alpha P_1 \longrightarrow TA \longrightarrow 0.$$

The transpose TA does not depend uniquely on A, but its derived objects $R^i_\beta(TA)$ for $i > 0$ are unique, up to isomorphism. That is, if $P_1' \longrightarrow P_o' \longrightarrow A \longrightarrow 0$ is another presentation of A with transpose $T'A$, then there are isomorphisms

$$R^i\beta(TA) \cong R^i\beta(T'A)$$

for all $i > 0$. But this can be made slightly more precise.

Lemma 3.1. Suppose TA and $T'A$ are transposes of A. Then there are morphisms $u\colon TA \longrightarrow T'A$ and $v\colon T'A \longrightarrow TA$ such that

$R^i\beta(uv) = 1$ <u>and</u> $R^i\beta(vu) = 1$ <u>for all</u> $i > 0$.

Proof. If $P_1 \longrightarrow P_o \longrightarrow A \longrightarrow 0$ and $P_1' \longrightarrow P_o' \longrightarrow A \longrightarrow 0$ are the presentations giving rise to TA and T'A respectively, there are complex morphisms $U\colon P_.' \longrightarrow P_.$ and $V\colon P_. \longrightarrow P_.'$ which extend $1\colon A \longrightarrow A$. The duals αV and αU induce the morphisms $v\colon T'A \longrightarrow TA$ and $u\colon TA \longrightarrow T'A$. Since the compositions $\alpha(V) \cdot \alpha(U)$ and $\alpha(U) \cdot \alpha(V)$ are homotopic to the identity the derived functors applied to $u \cdot v$ and $v \cdot u$ give the identity morphism. QED.

<u>Lemma</u> 3.2. <u>Every</u> B <u>in</u> $\underline{\underline{B}}$ <u>is the transpose of some</u> A <u>in</u> $\underline{\underline{A}}$.

Proof. Suppose $Q_1 \longrightarrow Q_o \longrightarrow B \longrightarrow 0$ is a presentation of B. Then the transpose TB has B as its transpose. That is $0 \longrightarrow \beta B \longrightarrow \beta Q_o \longrightarrow \beta Q_1 \longrightarrow TB \longrightarrow 0$ is exact. Apply α to get the exact sequence

$$0 \longrightarrow \alpha TB \longrightarrow \alpha\beta Q_1 \longrightarrow \alpha\beta Q_o$$

But $\alpha\beta Q_i = Q_i$ so the cokernel is B which is to say $B = TTB$. QED.

<u>Proposition</u> 3.3. <u>For each</u> F <u>in</u> $\underline{\underline{Coh}}[\underline{\underline{P}}^{op},\underline{\underline{Ab}}]$ <u>there is an exact sequence of functors</u>

$$0 \longrightarrow \mathrm{Ext}^1_{[\underline{\underline{P}},\underline{\underline{Ab}}]}(TF,-) \longrightarrow - \otimes F \longrightarrow$$

$$\mathrm{Hom}_{[\underline{\underline{P}},\underline{\underline{Ab}}]}(\alpha F,-) \longrightarrow \mathrm{Ext}^2_{[\underline{\underline{P}},\underline{\underline{Ab}}]}(TF,-) \longrightarrow 0$$

<u>natural in</u> F <u>and isomorphisms</u>

$$\mathrm{Ext}^{n+2}_{[\underline{\underline{P}},\underline{\underline{Ab}}]}(TA,-) \cong \mathrm{Ext}^n_{[\underline{\underline{P}},\underline{\underline{Ab}}]}(\alpha F,-) \quad \underline{\text{for all}} \quad n \geq 1. \qquad \text{QED.}$$

This result for rings is found in [Auslander, 2,3]. The proof of this general case is similar. (See also [Fossum, 20].)

<u>Corollary</u> 3.4. <u>If</u> E <u>is an injective object in</u> $[\underline{\underline{P}},\underline{\underline{Ab}}]$, <u>then</u>

$$E \otimes F \cong \mathrm{Hom}_{[\underline{\underline{P}},\underline{\underline{Ab}}]}(\alpha F,E)$$

<u>and consequently</u>

$$\mathrm{Tor}_n^{\underline{\underline{P}}}(E,F) \cong \mathrm{Hom}_{[\underline{\underline{P}},\underline{\underline{Ab}}]}(R^n\alpha F,E)$$

for all $n \geq 0$.

Proof. The first isomorphism follows from the proposition. The second follows from the fact that $\mathrm{Tor}_n^{\underline{\underline{P}}}$ can be computed by taking a projective resolution of F consisting of objects in $\underline{\underline{P}}$. Since E is injective, the functor $\mathrm{Hom}(-,E)$ commutes with homology. QED.

This is a standard isomorphism for (left) noetherian rings. That is if R is left noetherian the left module F is finitely generated, and E is right injective, then

$$\mathrm{Hom}_R(\mathrm{Hom}_R(.F,R),E) \cong E \otimes_R F.$$

Consequently $\mathrm{Tor}_n^R(E,F) \cong \mathrm{Hom}_R(\mathrm{Ext}_R^n(F,R),E)$ for all n. The same works for left coherent rings and F of finite presentation.

Proposition 3.5. Suppose P is projective in $\underline{\underline{\mathrm{Coh}}}[\underline{\underline{P}}^{op},\underline{\underline{Ab}}]$ and $L \in \underline{\underline{\mathrm{Coh}}}[\underline{\underline{P}},\underline{\underline{Ab}}]$. Then there are natural isomorphisms

$$\mathrm{Ext}^n_{[\underline{\underline{P}},\underline{\underline{Ab}}]}(L,\alpha P) \cong \mathrm{Hom}_{[\underline{\underline{P}}^{op},\underline{\underline{Ab}}]}(P,R^n\beta L) \quad \text{for all } n \geq 0.$$

Proof. Since P is projective, the functor $\mathrm{Hom}_{[\underline{\underline{P}}^{op},\underline{\underline{Ab}}]}(P,-)$ commutes with homology. Let $Q_. \longrightarrow L$ be a projective resolution of L by coherent projective objects in $[\underline{\underline{P}},\mathrm{Ab}]$. Then the homology of $Q_.$ is $R^{\cdot}\beta L$. But

$$\mathrm{Hom}_{[\underline{\underline{P}}^{op},\underline{\underline{Ab}}]}(P,\beta Q_.) \cong \mathrm{Hom}_{[\underline{\underline{P}},\underline{\underline{Ab}}]}(Q_.,\alpha P)$$

by the adjointness of α and β, and the homology of

$$\mathrm{Hom}_{[\underline{\underline{P}},\underline{\underline{Ab}}]}(Q_.,\alpha P) \quad \text{is} \quad \mathrm{Ext}^{\cdot}_{[\underline{\underline{P}},\underline{\underline{Ab}}]}(L,\alpha P). \qquad \text{QED.}$$

The remainder of this section is devoted to establishing a theorem due to M. Auslander which was presented in a course given at the University of Illinois in the fall semester of 1970. There is only one new feature, the flatness condition for the general pseudo duality. But even this is directly copied from Auslander's proof for coherent rings.

For the remainder of this section we denote by $\underline{\underline{A}}$ the category $\underline{\underline{\mathrm{Coh}}}[\underline{\underline{P}}^{op}, \underline{\underline{Ab}}]$, by $\underline{\underline{B}}$ the category $\underline{\underline{\mathrm{Coh}}}[\underline{\underline{P}}, \underline{\underline{Ab}}]$ and by $\underline{\underline{P}}$ and $\underline{\underline{Q}}$ the projectives in these categories, respectively.

Suppose n is a positive integer. An object A in $\underline{\underline{A}}$ is said to have *n-torsion* if $R^j\alpha A' = 0$ for all j in the range $0 \leq j < n$ and all (coherent) subobjects A' of A. The similar definition is made for $\underline{\underline{B}}$. The object A is said to have *grade* at least n, and we write grade $A \geq n$, if $R^i\alpha A = 0$ for $0 \leq j < n$. Thus A has n-torsion if and only if grade $A' \geq n$ for all subobjects A' of A in $\underline{\underline{A}}$.

Corollary 3.6. *An object* A *has* n*-torsion if and only if*

$$\mathrm{Ext}^j_{[\underline{\underline{P}}^{op}, \underline{\underline{Ab}}]}(A', P) = 0$$

for all j *with* $0 \leq j < n$, *for all subobjects* A' *of* A *and all projectives* P *in* $\underline{\underline{P}}$.

We now state Auslander's Theorem.

Auslander's Theorem 3.7. *The following statements are equivalent for the fixed integer* k.

a) *For all* A *in* $\underline{\underline{A}}$, *the derived objects* $R^i\alpha A$ *have* i-*torsion for* $1 \leq i \leq k$.

b) *For all* B *in* $\underline{\underline{B}}$, *the derived objects* $R^i\beta B$ *have* i-*torsion for* $1 \leq i \leq k$.

c) *For all* Q *in* $\underline{\underline{Q}}$, *if* $Q \longrightarrow I^{\cdot}$ *is a minimal injective resolution of* Q *in* $[\underline{\underline{P}}, \underline{\underline{Ab}}]$, $(=[\underline{\underline{Q}}^{op}, \underline{\underline{Ab}}])$ *then* flat dim $I^j \leq j$ *for* $0 \leq j < k$.

d) *For all* P *in* $\underline{\underline{P}}$, *if* $P \longrightarrow J^{\cdot}$ *is a minimal injective resolution of* P *in* $[\underline{\underline{P}}^{op}, \underline{\underline{Ab}}]$, *then* flat dim $J^j \leq j$ *for* $0 \leq j < k$.

We prove the equivalence of a) with c) (and thus, at the same time, the equivalence of b) with d)). Much later in this section we show the equivalence of a) with b).

Proof. Suppose condition a) is satisfied.

Suppose $Q \longrightarrow I^{\cdot}$ is a minimal injective resolution of Q in $[\underline{\underline{P}},\underline{\underline{Ab}}]$. To show that flat dim $I^j \leq j$, for $0 \leq j < k$, it is sufficient to show that $\mathrm{Tor}_i(I^j,A) = 0$ for all i with $i > j$ and all coherent A in $[\underline{\underline{P}}^{op},\underline{\underline{Ab}}]$. But

$$\mathrm{Tor}_i(I^j,A) \cong \mathrm{Hom}_{[\underline{\underline{P}},\underline{\underline{Ab}}]}(R^i\alpha A, I^j).$$

We show that $\mathrm{Hom}_{[\underline{\underline{P}},\underline{\underline{Ab}}]}(R^i\alpha A,I^j) = 0$ for $i > j$ by induction on j.

Suppose there is an A such that $\mathrm{Hom}_{[\underline{\underline{P}},\underline{\underline{Ab}}]}(R^1\alpha A,I^o) \neq 0$. Then there is a nonzero morphism $f\colon R^1\alpha A \longrightarrow I^o$. Let $Y = \mathrm{Imf} \cap Q$, which is nonzero since $Q \longrightarrow I^o$ is an essential extension and $\mathrm{Imf} \neq (0)$. There is thus a coherent subobject $B \subseteq R^1\alpha A$ such that the restriction of f to B, say $f|_B$, maps B to Y nontrivially. Thus $\mathrm{Hom}_{[\underline{\underline{P}},\underline{\underline{Ab}}]}(B,Q) \neq (0)$. But then $\beta B \neq (0)$. Since $R^1\alpha A$ is 1-torsion, this is a contradiction. So $\mathrm{Tor}_1(I^o,A) = 0$.

Our induction hypothesis is:
flat dim $I^j \leq j$ <u>for all</u> j <u>in the range</u> $0 \leq j < h - 1$ <u>for an integer</u> $n \leq k$.

We wish to show that flat dim $I^{n-1} \leq n - 1$.

If $\mathrm{Hom}_{[\underline{\underline{P}},\underline{\underline{Ab}}]}(R^n\alpha A,I^{n-1}) \neq (0)$ for some A, there is a coherent subobject $B \subseteq R^n\alpha A$ such that $\mathrm{Hom}_{[\underline{\underline{P}},\underline{\underline{Ab}}]}(B,W^{n-2}) \neq (0)$, where W^{n-2} is the $(n-2)^{th}$ cosyzygy of Q. (i.e. The sequences $0 \longrightarrow W^{n-2} \longrightarrow I^{n-1} \longrightarrow W^{n-1} \longrightarrow 0$ are exact.) Now $\mathrm{Ext}^j_{[\underline{\underline{P}},\underline{\underline{Ab}}]}(B,Q) \cong \mathrm{Hom}_{[\underline{\underline{P}}^{op},\underline{\underline{Ab}}]}(\beta Q,R^j\beta B)$. These groups are all zero, since $R^j\beta B = \overline{0}$ for $j \leq k$ by hypothesis a). Therefore the sequence

$$0 \longrightarrow \mathrm{Hom}(B,Q) \longrightarrow \mathrm{Hom}(B,I^o) \longrightarrow \ldots\ldots \longrightarrow \mathrm{Hom}(B,I^{n-2}) \longrightarrow$$

$$\mathrm{Hom}(B,W^{n-2}) \longrightarrow \mathrm{Ext}^{n-1}(B,Q) \longrightarrow 0$$

is exact. So $\mathrm{Hom}(B,I^{n-2}) \longrightarrow \mathrm{Hom}(B,W^{n-2})$ is a surjection. The remaining part of the sequence is exact by the induction hypothesis, so $\mathrm{Hom}(B,W^{n-2}) = 0$, a contradiction to the original assumption that $\mathrm{Hom}(R^n\alpha A,I^{n-1}) \neq 0$.

Suppose condition c) is satisfied. Then $\mathrm{Hom}_{[\underline{\underline{P}},\underline{\underline{Ab}}]}(R^n\alpha A, I^m) = \mathrm{Tor}_n(I^m, A) = 0$ for all $n > m$ and $m < k$. If B is a coherent subobject of $R^n\alpha A$, then

$$\mathrm{Hom}_{[\underline{\underline{P}},\underline{\underline{Ab}}]}(B, I^m) = 0.$$

But
$$\mathrm{Hom}_{[\underline{\underline{P}}^{op},\underline{\underline{Ab}}]}(\beta Q, R^m\beta B) \cong \mathrm{Ext}^m_{[\underline{\underline{P}},\underline{\underline{Ab}}]}(B,Q) .$$

The group $\mathrm{Ext}^m_{[\underline{\underline{P}},\underline{\underline{Ab}}]}(B,Q)$ is the mth homology of the complex $\mathrm{Hom}_{[\underline{\underline{P}},\underline{\underline{Ab}}]}(B,I^\cdot)$ which is zero for $m < n$. Hence $R^m\beta B = 0$, so $R^n\alpha A$ is $\overline{\overline{n}}$-torsion for all $1 \leqq n \leqq k$. This is condition a). QED.

In order to complete the proof of the theorem, two results concerning the pseudo-duality $\underline{\underline{A}} \underset{\beta}{\overset{\alpha}{\rightleftarrows}} \underline{\underline{B}}$ are needed.

<u>Proposition</u> 3.8. <u>Approximation Theorem</u> [Auslander and Bridger,5]. <u>Suppose an object</u> A <u>in</u> $\underline{\underline{A}}$ <u>has the property</u> $R^j\beta(R^i\alpha A) = 0$ <u>for all</u> $j < i$ <u>and all</u> i, $1 \leqq i \leqq k$. <u>Then there is an object</u> A_1 <u>and a homomorphism</u> $f: A \longrightarrow A_1$ <u>such that</u>

i) <u>the projective dimension</u> $\mathrm{pd}\, A_1 \leqq k$

<u>and</u> ii) <u>for each</u> i, $1 \leqq i \leqq k$, <u>the induced homomorphism</u> $R^i\alpha f$ <u>is a bijection</u>.

Proof. We go by induction on k. If $k = 1$, then $\beta(R^1\alpha A) = 0$. Let $Q_1 \longrightarrow Q_o \longrightarrow R^1\alpha A \longrightarrow 0$ be a presentation of $R^1\alpha A$ in $\underline{\underline{B}}$ by projective objects. Then $0 \longrightarrow \beta Q_o \longrightarrow \beta Q_1$ is exact in $\underline{\underline{A}}$.

Let A_1 be the cokernel, so

$$0 \longrightarrow \beta Q_o \longrightarrow \beta Q_1 \longrightarrow A_1 \longrightarrow 0$$

is exact. Thus $\mathrm{pd}\, A_1 \leqq 1$. Suppose

$$0 \longrightarrow Y \longrightarrow P_o \longrightarrow A \longrightarrow 0$$

is exact with P_o projective. Then

$$0 \longrightarrow \alpha A \longrightarrow \alpha P_o \longrightarrow \alpha Y \longrightarrow R^1\alpha A \longrightarrow 0$$

is exact in $\underline{\underline{B}}$ with αP_o projective. Also

$$0 \longrightarrow \alpha A_1 \longrightarrow Q_1 \longrightarrow Q_o \longrightarrow R^1\alpha A \longrightarrow 0$$

is exact. So there is a morphism of complexes

$$\begin{array}{ccccccccc} 0 \longrightarrow & \alpha A_1 & \longrightarrow & Q_1 & \longrightarrow & Q_o & \longrightarrow & R^1\alpha A & \longrightarrow 0 \\ & \downarrow f_2 & & \downarrow f_1 & & \downarrow f_o & & \downarrow = & \\ 0 \longrightarrow & \alpha A & \longrightarrow & \alpha P_o & \longrightarrow & \alpha Y & \longrightarrow & R^1\alpha A & \longrightarrow 0. \end{array}$$

Take β of this commutative diagram to obtain the diagram

$$\begin{array}{ccccccc} 0 \longrightarrow & \beta Q_o & \longrightarrow & \beta Q_1 & \longrightarrow & A_1 & \longrightarrow 0 \\ & \uparrow \beta f_o & & \uparrow \beta f_1 & & & \\ 0 \longrightarrow & Y & \longrightarrow & P_o & \longrightarrow & A & \longrightarrow 0 \end{array}$$

which induces $f\colon A \longrightarrow A_1$ with the desired property that $R^1\alpha f$ is a bijection.

There is a projective P in $\underline{\underline{A}}$ and a map $\pi\colon P \longrightarrow A_1$ such that $(\pi, f)\colon P \oplus A \longrightarrow A_1$ is an epimorphism. Let A_2 be the kernel so that $0 \longrightarrow A_2 \longrightarrow P \oplus A \longrightarrow A_1 \longrightarrow 0$ is exact. Then

$$0 \longrightarrow \alpha A_1 \longrightarrow \alpha(P \oplus A) \longrightarrow \alpha A_2 \longrightarrow R^1\alpha A \xrightarrow{R^1\alpha f} R^1\alpha A \longrightarrow R^1\alpha A_2$$
$$\longrightarrow R^2\alpha A_1 \longrightarrow \dots$$

is exact, the sequence

$$0 \longrightarrow \alpha A_1 \longrightarrow \alpha(P \oplus A) \longrightarrow \alpha A_2 \longrightarrow 0$$

is exact, the object $R^1\alpha A_2 = 0$ and

$$R^i\alpha A \longrightarrow R^i\alpha A_2$$

is an isomorphism for all $i > 1$ (since $\mathrm{pd}\, A_1 \leqq 1$).

Suppose $k > 1$. We can suppose we have found an object A' with

$\operatorname{pd} A' \leqq k - 1$ and an $f'\colon A \longrightarrow A'$ such that $R^i\alpha f'$ is an isomorphism for $1 \leqq i \leqq k - 1$. Let $\pi\colon P \longrightarrow A'$ be such that $(\pi, f')\colon P \oplus A \longrightarrow A'$ is an epimorphism with kernel A''. Then

$$0 \longrightarrow \alpha A' \longrightarrow \alpha(P \oplus A) \longrightarrow \alpha A'' \longrightarrow 0$$

is exact and the objects $R^i\alpha A'' = 0$ for $1 \leqq i \leqq k - 1$, while $R^k\alpha A \cong R^k\alpha A''$.

Let $Q_k \longrightarrow Q_{k-1} \longrightarrow \ldots \longrightarrow Q_1 \longrightarrow Q_o \longrightarrow R^k\alpha A \longrightarrow 0$ be exact. Then

$$0 \longrightarrow \beta Q_o \longrightarrow \ldots\ldots \longrightarrow \beta Q_k$$

is exact (since $R^i\beta(R^k\alpha A) = 0$ for $i < k$). Let A_1' be the cokernel of $\beta Q_{k-1} \longrightarrow \beta Q_k$. Let $0 \longrightarrow Y \longrightarrow P_{k-1} \longrightarrow \ldots \longrightarrow P_1 \longrightarrow P_o \longrightarrow A'' \longrightarrow 0$ be exact with each P_j projective. Then

$$0 \longrightarrow \alpha A'' \longrightarrow \alpha P_o \longrightarrow \alpha P_1 \longrightarrow \ldots \longrightarrow \alpha P_{k-1} \longrightarrow \alpha Y \longrightarrow R^k\alpha A \longrightarrow 0$$

is exact. There is a complex map $f_\bullet\colon Q_\bullet \longrightarrow \alpha P_\bullet$ which induces a morphism $f_1\colon A'' \longrightarrow A_1'$ such that $R^k\alpha f_1$ is an isomorphism. Let $\pi_1\colon P' \longrightarrow A_1'$ be a morphism such that $(\pi_1, f_1)\colon P' \oplus A'' \longrightarrow A_1'$ is an epimorphism with kernel A_2. We get the diagram with exact rows and columns which defines A_1:

$$\begin{array}{ccccccccc}
 & & 0 & & 0 & & & & \\
 & & \downarrow & & \downarrow & & & & \\
 & & A_2 & \xrightarrow{=} & A_2 & & & & \\
 & & \downarrow & & \downarrow & & & & \\
0 & \longrightarrow & P' \oplus A'' & \longrightarrow & P' \oplus (P \oplus A) & \longrightarrow & A' & \longrightarrow & 0 \\
 & & \downarrow & & \downarrow & & \downarrow{=} & & \\
0 & \longrightarrow & A_1' & \longrightarrow & A_1 & \longrightarrow & A' & \longrightarrow & 0. \\
 & & \downarrow & & \downarrow & & & & \\
 & & 0 & & 0 & & & &
\end{array}$$

Since $\operatorname{pd} A' \leqq k - 1$ and $\operatorname{pd} A_1' \leqq k$, we get $\operatorname{pd} A_1 \leqq k$. The object A_1 satisfies the properties and

$$0 \longrightarrow \alpha A_1 \longrightarrow \alpha(P_1 \oplus P \oplus A) \longrightarrow \alpha A \longrightarrow 0 \quad \text{is exact.} \qquad \text{QED.}$$

The same notations and indeed the same ideas are used to establish the next result.

Lemma 3.9. *Suppose* A *in* $\underline{\underline{A}}$ *has the property* $R^j\beta(R^i\alpha A) = 0$ *for all* $0 \leq j \leq i$ *and all* i *with* $1 \leq i \leq k$. *If* $\alpha A = 0$, *then* $R^i\alpha A = 0$ *for* $0 \leq i \leq k$.

Proof. We can suppose, by induction, that $R^i\alpha A = 0$ for $0 \leq i < k$. Let $f\colon A \longrightarrow A_1$ satisfy the conditions of Proposition 3.8. Suppose $\pi\colon P \longrightarrow A_1$ is a morphism such that $(\pi,f)\colon P \oplus A \longrightarrow A_1$ is an epimorphism, with P a projective object. Suppose A_2 is the kernel so that $0 \longrightarrow A_2 \longrightarrow P \oplus A \longrightarrow A_1 \longrightarrow 0$ is exact. Then also

$$0 \longrightarrow \alpha A_1 \longrightarrow \alpha(P \oplus A) \longrightarrow \alpha A_2 \longrightarrow 0$$

is exact. Hence we get the commutative diagram with exact rows

$$\begin{array}{ccccccccc}
 & & & & 0 & & & & \\
 & & & & \downarrow & & & & \\
 & & & & A & \overset{f}{\searrow} & & & \\
 & & & & \downarrow & & & & \\
0 & \longrightarrow & A_2 & \longrightarrow & P \oplus A & \longrightarrow & A_1 & \longrightarrow & 0 \\
 & & \downarrow & & \downarrow & & \downarrow & & \\
0 & \longrightarrow & \beta\alpha A_2 & \longrightarrow & P & \longrightarrow & \beta\alpha A_1 & . &
\end{array}$$

But $A_1 \longrightarrow \beta\alpha A_1$ is a monomorphism since $R^k\beta(R^k\alpha(A)) = 0$. Hence f, which is the composition $A \longrightarrow P \oplus A \longrightarrow A_1$, factors through $A_2 \longrightarrow P \oplus A$. That is, the morphism $f = 0$. Hence $R^k\alpha f = 0$. But $R^k\alpha f$ is an isomorphism, so $R^k\alpha A = 0$. QED.

We can now complete the proof of Auslander's theorem. We will show that $R^i\alpha A$ has i-torsion for all i with $1 \leq i \leq k$ implies that $R^i\beta B$ has i-torsion for i in the same range. This will show that a) implies b). The same proof shows that b) implies a).

We go by induction on k. Suppose $R^1\alpha A$ has 1-torsion for all A in $\underline{\underline{A}}$, and suppose $A' \subsetneqq R^1\beta B$ for some B in B. We want to show that $\alpha A' = 0$.

Note that $B = T(T(B))$. Hence we have the exact sequence

$$0 \to R^1\beta B \to TB \to B\alpha TB \to R^2\beta B \to 0$$

by Proposition 3.3. Let X denote the cokernel $R^1\beta B \to TB$. We get two exact sequences

$$0 \to \alpha R^2\beta B \to \alpha\beta\alpha TB \to \alpha X \to R^1\alpha R^2\beta B \to \dots$$

and

$$0 \to \alpha X \to \alpha TB \to \alpha R^1\beta B \to R^1\alpha X \to \dots \quad .$$

Now $\alpha\beta\alpha A \to \alpha A$ is a split epimorphism since the composition $\alpha A \to \alpha\beta\alpha A \to \alpha A$ is the identity. Hence the morphism $\alpha X \to \alpha TB$ is an isomorphism. Therefore $\alpha(R^1\beta B)$ is isomorphic to a subobject of $R^1\alpha X$. Hence $\beta\alpha(R^1\beta B) = 0$. But then $\alpha R^1\beta B = 0$.

Now suppose A' is a subobject of $R^1\beta B$, and let A'' be the cokernel of the composition $A' \to R^1\beta B \to TB$. There is induced an exact sequence $0 \to A' \to R^1\beta B \to A'' \to X \to 0$. Since $\alpha R^1\beta B = 0$, the morphism $\alpha X \to \alpha A''$ is an isomorphism. But $\alpha X \to \alpha TB$ is also an isomorphism. Therefore $\alpha A'' \to \alpha TB$ is an isomorphism. Hence $\alpha A'$ is isomorphic to a subobject of $R^1\alpha A''$. Since $R^1\alpha A''$ has 1-torsion, the object $\beta\alpha A' = 0$ and so $\alpha A' = 0$. Hence $R^1\beta B$ has 1-torsion.

Suppose that $k > 1$ and that we have shown that $R^i\beta B$ has i-torsion for $1 \leq i \leq k - 1$.

We need only show that $R^k\beta B$ has k-torsion in order to verify condition b). Since $k > 1$, each object $R^k\beta B \cong R^{k-1}\beta B$, for some B' in $\underline{\underline{B}}$. Hence $R^k\beta B$ has $k - 1$ torsion by the induction hypothesis. We may suppose $R^j\beta B = 0$ for $0 < j < k$ by Proposition 3.8. Let

$$Q_k \to \dots \to Q_o \to B \to 0$$

be a projective resolution of B. Then

$$0 \to \beta B \to \beta Q_o \to \dots \to \beta Q_k$$

is exact. Let T be the cokernel of $\beta Q_{k-1} \to \beta Q_k$. Then $R^i\alpha T = 0$ for $o < i < k$. Furthermore T contains a subobject isomorphic to $R^k\beta B$. Let Y be a subobject of $R^k\beta B$ and consider it as a subobject of T. Since $R^{k-1}\alpha T = 0$, we get induced a monomorphism

$$R^{k-1}\alpha Y \longrightarrow R^k\alpha(T/Y).$$

We know, by induction, that $R^j\alpha Y = 0$ for $j < k - 1$. Since $R^{k-1}\alpha Y$ is a subobject of $R^k\alpha(T/Y)$, the object $R^j\beta(R^{k-1}\alpha Y) = 0$ for $0 \leq j < k$. By Lemma 3.9, the object $R^{k-1}\alpha Y = 0$. Therefore $R^k\beta B$ is k-torsion. This completes the proof of Auslander's Theorem. QED.

Definition. Suppose $\underline{\underline{A}} \underset{\beta}{\overset{\alpha}{\rightleftarrows}} \underline{\underline{B}}$ is a pseudo-duality, and k is a positive integer. We say $\underline{\underline{A}}$ (and thus also $\underline{\underline{B}}$) is k-*Gorenstein* if one of the equivalent conditions of Auslander's Theorem is satisfied. If $\underline{\underline{A}}$ is k-Gorenstein for all k, then $\underline{\underline{A}}$ is *Gorenstein*.

This concept is used to generalize the concept of a Gorenstein ring for commutative rings to the class of not necessarily commutative rings. Thus, for example, if A is a Gorenstein ring, then the category of finitely generated A-modules, $\underline{\underline{\mathrm{Mod}}}_A^f$, is in pseudo-duality with itself by the functor $\alpha = \mathrm{Hom}_A(-,A)$ and is Gorenstein.

More generally, if A is a noetherian commutative ring and $\underline{\underline{A}}$ is the category of finitely generated A-modules, then $\mathrm{Hom}_A(-,A)$ establishes a pseudo-duality on $\underline{\underline{A}}$. In [Fossum and Reiten, 21] there are many conditions which are equivalent to the condition that $\underline{\underline{A}}$ be k-Gorenstein. One of these is the local condition: If $\underline{p} \in \mathrm{Spec}\, A$ and depth $A_{\underline{p}} < k$, then $A_{\underline{p}}$ is Gorenstein. Thus every integral domain is 1-Gorenstein (in fact every reduced ring is 1-Gorenstein). Every normal ring is 2-Gorenstein by Serre's criteria [EGA, 29].

In section 5, we will see many applications of the theory of Gorenstein rings. But one of our main goals is to show that k-Gorenstein rings are sufficiently rich. In order to do this we now demonstrate one of the main results of this paper.

Theorem 3.10. *Suppose* A *is left and right coherent. Then the ring* $\begin{pmatrix} A & 0 \\ A & A \end{pmatrix}$ *is left and right coherent. Furthermore it is* k-*Gorenstein if and only if* A *is* k-*Gorenstein*.

Proof. That $\begin{pmatrix} A & 0 \\ A & A \end{pmatrix}$ is left and right coherent follows from Roos' result which is our Corollary 2.3.

In order to show the Gorenstein properties we need to construct a minimal resolution of the ring $\begin{pmatrix} A & 0 \\ A & A \end{pmatrix}$. Denote this ring by B. Returning to former principles, we note that a left B-module is just a pair of A-modules $\binom{M}{N}_f$ and a homomorphism $f: M \longrightarrow N$. Then an element

$$\begin{pmatrix} a & 0 \\ b & c \end{pmatrix} \binom{m}{n} = \binom{am}{f(mb) + cn}.$$

Using our results concerning injective objects from the first section, we know that $\binom{M}{N}_f$ is injective if and only if ker f is injective, the module N is injective, and f is a surjection. In general, if E denotes the injective envelope of ker f and F denotes the injective envelope of N, then the injective envelope of $\binom{M}{N}_f$ is $\binom{E \oplus F}{F}_{(0,1)}$

Now B is the direct sum of the two left ideals $\binom{A}{A}$ and $\binom{0}{A}$. To compute a minimal injective resolution of B as left modules, it is sufficient to compute a minimal injective resolution of $\binom{A}{A}$ and $\binom{0}{A}$ and then paste them together. Let $0 \longrightarrow A \longrightarrow E^0 \longrightarrow E^1 \longrightarrow \ldots$ be a minimal injective resolution of the left A-module A. Let the left modules W^i be defined inductively: First $0 \longrightarrow A \longrightarrow E^0 \longrightarrow W^0 \longrightarrow 0$ is exact and for $i > 0$, the sequences $0 \longrightarrow W^{i-1} \longrightarrow E^i \longrightarrow W^i \longrightarrow 0$ are exact. Let $\underline{\pi}^i : E^i \longrightarrow W^i$ be the projections.

It is clear that the left B-modules $\binom{E^i}{E^i}$ are injective and that

$$0 \longrightarrow \binom{A}{A} \longrightarrow \binom{E^0}{E^0} \longrightarrow \binom{E^1}{E^1} \longrightarrow \ldots$$

is a minimal injective resolution of $\binom{A}{A}$. Now the injective envelope of $\binom{0}{A}$ is $\binom{E^0}{E^0}$. The cokernel of $\binom{0}{A} \longrightarrow \binom{E^0}{E^0}$ is $\binom{E^0}{W^0}$.

The injective envelope of $\binom{E^0}{W^0}$ is $\binom{E^0 \oplus E^1}{E^1}_{(0,1)}$.

The cokernel of $\binom{E^0}{W^0} \longrightarrow \binom{E^0 \oplus E^1}{E^1}$ is $\binom{E^1}{W^1}_{\underline{\pi}^1}$.

Thus, inductively, we see that the ith term in a minimal injective resolution of

$$\binom{0}{A} \text{ is } \binom{E^{i-1} \oplus E^i}{E^i}_{(0,1)} \qquad (\text{where } E^{-1} = (0)).$$

Thus, the injective envelope of $\begin{pmatrix} A & 0 \\ A & A \end{pmatrix}$ is $\begin{pmatrix} E^0 & E^0 \\ E^0 & E^0 \end{pmatrix}$ while the ith term in an injective resolution is

$$\begin{pmatrix} E^i & E^{i-1}\oplus E^i \\ E^i & E^i \end{pmatrix} \quad \text{for} \quad i > 0.$$

Let I^i denote this i^{th} term. Then $I^0 = \binom{E^0}{E^0}^2$ while for $i > 0$

$$I^i = \binom{E^i}{E^i}^2 \oplus \binom{E^{i-1}}{0}.$$

Thus we can compute the flat dimension of I^i. In fact

$$\text{flat dim}_B I^0 = \text{flat dim}_B \binom{E^0}{E^0}$$

while $\text{flat dim}_B I^i = \sup\left\{ \text{flat dim}_B \binom{E^i}{E^i}, \text{flat dim}_B \binom{E^{i-1}}{0} \right\}$.

Moreover, we have an exact sequence

$$0 \longrightarrow \binom{0}{E^i} \longrightarrow \binom{E^i}{E^i} \longrightarrow \binom{E^i}{0} \longrightarrow 0$$

of left B-modules. Now we know that the functor $X \longrightarrow \binom{X}{X}_{id}$ is exact. Also a left B-module $\binom{M}{N}_f$ is flat if and only if M is flat, f is an injection and coker f is flat. Therefore

$$\text{flat dim}_B \binom{X}{X} = \text{flat dim}_A X .$$

But also the functor $X \longrightarrow \binom{0}{X}$ is exact

and $$\text{flat dim}_B \binom{0}{X} = \text{flat dim}_A X .$$

Therefore, we have $\text{flat dim}_B \binom{X}{0} = 1 + \text{flat dim}_A X$.

Hence $$\text{flat dim}_B I^0 = \text{flat dim}_A E^0$$

and for $i > 0$, $\text{flat dim}_B I^i = \sup\{\text{flat dim}_A E^i, 1 + \text{flat dim}_A E^{i-1}\}$.

Now we suppose k is an integer, $k \geq 1$. Then $\text{flat dim}_B I^i \leq i$ for all $i < k$ if and only if $\text{flat dim}_A E^i \leq i$ for all $i < k$. Hence A is k-Gorenstein if and only if B is k-Gorenstein. QED.

To see exactly how touchy the Gorenstein property is, we note that the ring $\begin{pmatrix} A & 0 \\ A\oplus A & A \end{pmatrix}$ is not 1-Gorenstein. For the injective envelope of $\binom{A}{A\oplus A}$ is not what one would expect. Indeed it is the left module

$$I = \binom{\text{Hom}_A(A\oplus A,\ E^0\oplus E^0)}{E^0\oplus E^0}_{\tau}$$

with the natural trace homomorphism

$$\tau : (A\oplus A) \otimes_A \mathrm{Hom}(A\oplus A,\ E^0\oplus E^0) \longrightarrow E^0\oplus E^0 ,$$

and where

$$A \longrightarrow \mathrm{Hom}_A(A\oplus A,\ E^0\oplus E^0)$$

is the composition $A \longrightarrow \mathrm{End}_A(A\oplus A) \longrightarrow \mathrm{Hom}_A(A\oplus A,\ E^0\oplus E^0)$.
The left module I is flat if and only if τ is an injection, coker τ is flat and $\mathrm{Hom}(A\oplus A,\ E^0\oplus E^0)$ is flat (Prop. 1.14 (bis)). But τ is clearly not an injection, and hence I is not flat.

Thus even though the bimodule M is of a very nice sort, the ring $\begin{pmatrix} A & 0 \\ M & A \end{pmatrix}$ need not be 1-Gorenstein.

One very general question which we have not been able to solve is: What are necessary and sufficient conditions in order that the ring $A \ltimes M$ be (k-) Gorenstein?

We have one very special case where some sort of answer can be given.

Proposition 3.11. Suppose M is an A-bimodule which is right flat. If $A \ltimes M$ is k-Gorenstein, then A is k-Gorenstein.

Proof. From the techniques and the result, Proposition 1.14 (bis), we know that the minimal injective resolution of $A \ltimes M$ is a complex of the form

$$\begin{matrix} FGI^\bullet \oplus FI^\bullet \\ \Big\downarrow \begin{pmatrix} 0 & 0 \\ 1 & 0 \end{pmatrix} \\ GI^\bullet \oplus I^\bullet \end{matrix}$$

where $I^\bullet$ is a complex of injective left A-modules, the functor $F = M\otimes_A$ and $G = \mathrm{Hom}_A(.M.,-)$. Furthermore, since

$$\text{flat dim} \left(\begin{matrix} FGI^i \oplus FI^i \\ \\ GI^i \oplus I^i \end{matrix} \right) \leq i \quad \text{for} \quad i < k,$$

we get

$$\text{flat dim}_A\ GI^i \leq i \quad \text{for} \quad i < k.$$

But $GI^\bullet$ is an injective complex since M is right flat. There is a minimal injective resolution $J^\bullet$ of A embedded in the complex $GI^\bullet$, so each J^i is a direct summand of GI^j. Hence flat $\dim_A J^i \leq i$ for

$i < k$. Thus A is k-Gorenstein. QED.

We leave to the reader the interpretation of this result for triangular matrix rings. In Section 5 we answer the question for A Noetherian and M of finite type.

Section 4. Homological dimensions in $\underline{\underline{A}} \ltimes F$

Throughout this section, we shall use (generally without reference) the notation and conventions as described in Section 1. In particular, all categories are assumed to be abelian with enough projectives (enough injectives, if injective dimension is being considered).

Our first (expository) paper [22] on this subject dealt with finitistic projective dimension, as defined by Auslander and Buchsbaum [6] and further studied by Bass [10,11] only for the special trivial extension category $\underline{\underline{\text{Map}}}(F\ \underline{\underline{A}},\ \underline{\underline{B}})$. Our results were especially precise when the functor $F: \underline{\underline{A}} \longrightarrow \underline{\underline{B}}$ was exact and, in any case, they generalized the corresponding results of Chase [15], Eilenberg-Rosenberg- Zelinsky [17], Harada [31], Mitchell [41,42,43] and Fields [19] on the global dimension of trianglular matrix rings. Since the time of the writing of our first paper, Palmer and Roos [56,57] have made nearly a complete determination of the situation in which left gl. dim $(R \ltimes M) < \infty$. In particular, the Palmer-Roos calculation of left gl. dim $(R \ltimes M)$ (when left gl. dim $(R \ltimes M) < \infty$) is more precise than our calculations [22] and Reiten's estimates [62].

The technique of Palmer and Roos [57] is that of spectral sequences and their results are most satisfactory when $R \ltimes M$ is a triangular matrix ring or when M is flat as a left or right R-module. In part A of this section, we shall add our own result (Theorem 4.14) on the finiteness of gl. dim $(\underline{\underline{A}} \ltimes F)$ which, presumably, is not a trivial consequence of the work in [56,57]. However, we point out that it is quite often the case that gl dim$(\underline{\underline{A}} \ltimes F) = \infty$ (even where gl. dim $\underline{\underline{A}} < \infty$). For example, if R is a commutative Noetherian ring and if $M \neq 0$ is a finitely generated (symmetric) R-module, then it is elementary to show gl. dim $(R \ltimes M) = \infty$. Nevertheless, $R \ltimes M$ may possess various non-trivial modules having either finite projective or injective dimension which provide a great deal of information about R and M. This is precisely the situation in Section 5 (Gorenstein Modules). Thus, we shall mainly concern ourselves with objects in $\underline{\underline{A}} \ltimes F$ (respectively

$G \rtimes \underline{\underline{A}}$) having finite projective (respectively, injective) dimension. Our investigation leads to a calculation of the $FPD(\underline{\underline{A}} \ltimes F)$, that is, the finitistic projective dimension of $\underline{\underline{A}} \ltimes F$, in many situations. In addition, we shall provide ample examples of the (sometimes pathological) behavior of $FDP(\underline{\underline{A}} \ltimes F)$.

The organization of this section is as follows:

A. General remarks on projective dimension in $\underline{\underline{A}} \ltimes F$ (injective dimension in $G \rtimes \underline{\underline{A}}$).

B. The finitistic projective dimension of $Map(F\ \underline{\underline{A}}, \underline{\underline{B}})$ and triangular matrix rings (Results to be applied in Sections 6 and 7).

C. The finitistic projective dimension of $R \ltimes M$ when R is a commutative ring and M is a symmetric R-bimodule.

D. The injective dimension of $R \ltimes M$ as a (left) module. (Results to be applied in Sections 5 and 6).

4. A. General remarks on projective dimension in $\underline{\underline{A}} \ltimes F$.

Let $\underline{\underline{A}}$ be an abelian category with enough projectives. Then $pd_{\underline{\underline{A}}}A$ (respectively, $id_{\underline{\underline{A}}}A$, if $\underline{\underline{A}}$ has enough injectives) denotes the projective dimension (respectively, injective dimension) of the object A in $\underline{\underline{A}}$ and $FPD(\underline{\underline{A}}) = \sup\{pd_{\underline{\underline{A}}}A : pd_{\underline{\underline{A}}}A < \infty\}$.

Our first lemma deals with objects in $\underline{\underline{A}} \times F$ of the general form

$$\begin{array}{c} FB \oplus FD \\ \searrow \ \downarrow \\ B \oplus D \end{array} ,$$

for example

$$ZA = \begin{array}{c} FA \\ \downarrow 0 \\ A \end{array} \qquad \text{or} \qquad TA = \begin{array}{c} FA \oplus F^2A \\ \searrow = \\ A \oplus FA \end{array} .$$

Lemma 4.1. *If* α *represents an object in* $\underline{\underline{A}} \ltimes F$ *such that* codomain $\alpha = B \oplus D$ *with* image $\alpha \subseteq D$, *then*

$$pd_{\underline{\underline{A}}}B \leq pd_{\underline{\underline{A}} \ltimes F}(\alpha).$$

<u>In particular, if</u> A <u>is in</u> $\underline{\underline{A}}$, <u>then</u>

$$pd_{\underline{\underline{A}}}A \leq \min(pd_{\underline{\underline{A}} \ltimes F}ZA,\ pd_{\underline{\underline{A}} \ltimes F}TA).$$

Proof. We may assume that $pd_{\underline{\underline{A}} \ltimes F}(\alpha) = n < \infty$. If $n = 0$, the conclusion follows easily from the structure of projective objects in $\underline{\underline{A}} \ltimes F$ (See Corollary 1.6.). Therefore, we proceed by induction on n $(n \geq 1)$.

Let $0 \longrightarrow K \longrightarrow P \overset{\epsilon}{\longrightarrow} B \longrightarrow 0$ be exact in $\underline{\underline{A}}$ with P projective, and let $Q \overset{\delta}{\longrightarrow} D$ be an epimorphism in $\underline{\underline{A}}$ with Q projective. We now recall that α has the form

$$\begin{array}{ccc} FB & \oplus & FD \\ & {\scriptstyle\alpha'}\searrow & \downarrow {\scriptstyle\alpha''} \\ B & \oplus & D. \end{array}$$

If L is the kernel of the morphism

$$\lambda = (\alpha' F\varepsilon,\ \delta,\ \alpha'' F\delta) : FP \oplus Q \oplus FQ \longrightarrow D$$

and if

$$\mu = \begin{pmatrix} \epsilon & 0 \\ 0 & \lambda \end{pmatrix} : P \oplus (FP \oplus Q \oplus FQ) \longrightarrow B \oplus D,$$

then a standard diagram chase shows that we obtain an exact sequence in $\underline{\underline{A}} \ltimes F$ of the form

$$\begin{array}{ccccccc} & FK \oplus FL & \longrightarrow & FP \oplus F^2P \oplus FQ \oplus F^2Q & \longrightarrow & FB \oplus FD & \longrightarrow 0 \\ & \searrow\ \downarrow & & \searrow_{=} \qquad \searrow_{=} & & {\scriptstyle\alpha'}\searrow\ \downarrow{\scriptstyle\alpha''} & \\ 0 \longrightarrow & K \oplus L & \longrightarrow & P \oplus FP \oplus Q \oplus FQ & \underset{\mu}{\longrightarrow} & B \oplus D & \longrightarrow 0, \end{array}$$

where the middle term represents a projective object in $\underline{\underline{A}} \ltimes F$. Hence, by induction, we conclude

$$pd_{\underline{\underline{A}} \ltimes F}(\alpha) = 1 + pd_{\underline{\underline{A}} \ltimes F}(\beta) \geq 1 + pd_{\underline{\underline{A}}}K = pd_{\underline{\underline{A}}}B,$$

where β denotes the object $\begin{array}{ccc} FK & \oplus & FL \\ & \searrow & \downarrow \\ K & \oplus & L. \end{array}$

QED.

The next lemma is merely the dual of Lemma 4.1 for the category $G \rtimes \underline{\underline{A}}$, where G is a left exact functor (See Section 1). Its statement concerns objects in $G \rtimes \underline{\underline{A}}$ of the form

$$\begin{array}{ccc} D & \oplus & B \\ \downarrow & \searrow & \\ GD & \oplus & GB \end{array}$$

for example $ZA = \begin{array}{c} A \\ \downarrow\ 0 \\ GA \end{array}$ and $HA = \begin{array}{ccc} GA & \oplus & A \\ & \searrow & = \\ G^2A & \oplus & GA \end{array}$.

Lemma 4.2. *If* β *represents an object in* $G \rtimes \underline{\underline{A}}$ *such that* domain $\beta = D \oplus B$ *with* $B \subseteq \ker \beta$, *then*

$$\mathrm{id}_{\underline{\underline{A}}} B \leq \mathrm{id}_{G \rtimes \underline{\underline{A}}}(\beta).$$

In particular, if $A \in \underline{\underline{A}}$, *then*

$$\mathrm{id}_{\underline{\underline{A}}} A \leq \min\left(\mathrm{id}_{G \rtimes \underline{\underline{A}}} ZA,\ \mathrm{id}_{G \rtimes \underline{\underline{A}}} HA\right). \qquad \text{QED.}$$

In general, we shall not explicitly state theorems for $G \rtimes \underline{\underline{A}}$ which are dual to those in $\underline{\underline{A}} \ltimes F$ except in cases where the dual statement will be needed for later application.

Corollary 4.3. gl. dim $\underline{\underline{A}} \leq$ gl. dim$(\underline{\underline{A}} \ltimes F)$.

We shall demonstrate in Part C (Example 4.30) of this section that Corollary 4.3 does not remain true when one replaces global dimension by finitistic projective dimension.

Lemma 4.4. *If* A *is an object in* $\underline{\underline{A}}$ *such that* $L_iF(A) = 0$ *for all* $i > 0$ (L_iF *is the* i^{th} *left derived functor of* F), *then*

$$\mathrm{pd}_{\underline{\underline{A}} \ltimes F}(TA) = \mathrm{pd}_{\underline{\underline{A}}} A.$$

Dually, if A *is an object in* $\underline{\underline{A}}$ *such that* $R^iG(A) = 0$ *for all* $i > 0$, *then*

$$\mathrm{id}_{G \rtimes \underline{\underline{A}}}(HA) = \mathrm{id}_{\underline{\underline{A}}} A.$$

Proof. The hypothesis $L_iF(A) = 0$, for $i > 0$, simply allows us to "lift" (via the functor T) a projective resolution of A in $\underline{\underline{A}}$ to a projective resolution of TA in $\underline{\underline{A}} \ltimes F$. Hence,

$$pd_{\underline{\underline{A}} \ltimes F}(TA) \leq Pd_{\underline{\underline{A}}}A.$$

The reverse inequality is a consequence of Lemma 4.1. QED.

In the remainder of this section, the cokernel functor $C: \underline{\underline{A}} \ltimes F \to \underline{\underline{A}}$ as defined in Section 1, plays a vital role in determining the projective dimension of objects in $\underline{\underline{A}} \ltimes F$. Especially, we shall be concerned with the vanishing of the higher left derived functors of C. The advantage of such an occurrence can be observed in Theorem 4.12. However, before we discuss further projective dimension in $\underline{\underline{A}} \ltimes F$, we need some additional machinery in order to draw a closer connection between the properties of the functors $F: \underline{\underline{A}} \to \underline{\underline{A}}$ and $C: \underline{\underline{A}} \ltimes F \to \underline{\underline{A}}$.

Let $\alpha : FA \to A$ represent an object in $\underline{\underline{A}} \ltimes F$. Since $\alpha F\alpha = 0$, it follows that $F^n(\alpha)\, F^{n+1}(\alpha) = 0$, for $n > 0$. Hence, we obtain a complex

$$\cdots \to F^3A \xrightarrow{F^2\alpha} F^2A \xrightarrow{F\alpha} FA \xrightarrow{\alpha} A,$$

which we call the associated F-complex over α and denote it (when necessary) by $\underline{\underline{C}}^F(\alpha)$. The homology modules of this complex will be denoted by $H_i^F(\alpha)$ and α will be called an acyclic object if $H_i^F(\alpha) = 0$, for all $i > 0$. We note that $H_o^F(\alpha) = \text{cod}\ \alpha = C(\alpha)$.

4.5. <u>Observations concerning the</u> $H_i^F(\alpha)$.

1) If A is in $\underline{\underline{A}}$, then $H_i^F(TA) = 0$ for $i > 0$. Also $H_i^F(ZA) = F^iA$ for all $i > 0$.

2) Let $\alpha : FA \to A$ be an object in $\underline{\underline{A}} \ltimes F$ and let $\pi : A \to \text{cok}\ \alpha$ denote the cokernel of α. Since right exact functors preserve cokernels, we obtain, for each $n \geq 0$, a commutative triangle

$$\begin{array}{ccc} F^{n+1}A & \xrightarrow{F^{n+1}\pi} & \\ {\scriptstyle F^n_\alpha}\downarrow & & F^{n+1}(\text{cok}\ \alpha) \\ F^nA & \xrightarrow{\lambda_n} & \end{array}$$

where $F^{n+1}\pi$ is (necessarily) an epimorphism and λ_n is the unique morphism such that $F^n\alpha = \lambda_n F^{n+1}\pi$. Moreover, there is a natural isomorphism $H_{n+1}^F(\alpha) \cong \text{Ker}\ \lambda_n$.

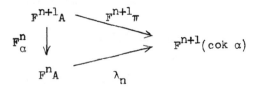

The exact sequence

$$\begin{array}{ccccccccc}
 & & FC & \longrightarrow & FB & \longrightarrow & FA & \longrightarrow & 0 \\
 & & \downarrow{\scriptstyle \gamma} & & \downarrow{\scriptstyle \beta} & & \downarrow{\scriptstyle \alpha} & & \\
0 & \longrightarrow & C & \xrightarrow{i} & B & \xrightarrow{j} & A & \longrightarrow & 0
\end{array}$$

in $\underline{\underline{A}} \ltimes F$ gives a commutative diagram with exact rows

$$\begin{array}{ccccccccc}
 & & F(\mathrm{cok}\ \gamma) & \longrightarrow & F(\mathrm{cok}\ \beta) & \longrightarrow & F(\mathrm{cok}\ \alpha) & \longrightarrow & 0 \\
 & & \downarrow & & \downarrow & & \downarrow & & \\
0 & \longrightarrow & C & \xrightarrow{i} & B & \xrightarrow{j} & A & \longrightarrow & 0
\end{array}$$

where the top row is obtained by applying the functor F to the exact sequence $\mathrm{cok}\ \gamma \longrightarrow \mathrm{cok}\ \beta \longrightarrow \mathrm{cok}\ \alpha \longrightarrow 0$.

3) Again let $\alpha : FA \longrightarrow A$ represent an arbitrary object in $\underline{\underline{A}} \ltimes F$. We obtain a third quadrant double complex (to the left of and below the black line).

$$\begin{array}{ccccccccc|ccccc}
\cdots & \longrightarrow & P_2 & \longrightarrow & P_1 & \longrightarrow & P_0 & & & \longrightarrow & \mathrm{cok}\ \alpha & \longrightarrow & 0 \\
\hline
 & & \uparrow{\scriptstyle =} & & \nearrow{\scriptstyle =} & & \nearrow{\scriptstyle =} & & & & \uparrow & & \\
\cdots & \longrightarrow & P_2 \oplus FP_2 & \longrightarrow & P_1 \oplus FP_1 & \longrightarrow & P_0 \oplus FP_0 & & & \longrightarrow & A & \longrightarrow & 0 \\
 & & \nearrow{\scriptstyle =} & & \nearrow{\scriptstyle =} & & \nearrow{\scriptstyle =} & & & & \uparrow & & \\
\cdots & \longrightarrow & FP_2 \oplus F^2P_2 & \longrightarrow & FP_1 \oplus F^2P_1 & \longrightarrow & FP_0 \oplus F^2P_0 & & & \longrightarrow & FA & \longrightarrow & 0 \\
 & & \nearrow{\scriptstyle =} & & \nearrow{\scriptstyle =} & & \nearrow{\scriptstyle =} & & & & \uparrow & & \\
\cdots & \longrightarrow & F^2P_2 \oplus F^3P_2 & \longrightarrow & F^2P_1 \oplus F^3P_1 & \longrightarrow & F^2P_0 \oplus F^3P_0 & & & \longrightarrow & F^2A & \longrightarrow & 0
\end{array}$$

The double complex above has the following properties:

3a) The first row is a result of an $\underline{\underline{A}} \ltimes F$ projective resolution of α and hence is $\underline{\underline{A}}$-exact. The remaining rows are obtained by successive application of the functor F to the first row.

3b) The columns are split exact. In particular, the $(-i)^{th}$ column is the associated F-complex over the $\underline{\underline{A}} \ltimes F$-projective $T(P_i)$. Taking the zeroeth homology along the rows gives the associated F-complex over α (as indicated to the right of the black line in the above diagram).

3c) Taking the zeroeth homology along the columns gives a projective complex in $\underline{\underline{A}}$ whose zeroeth homology is $\mathrm{cok}\ \alpha$ and, in general whose $n^{\underline{th}}$ homology is $L_nC(\alpha)$.

3d) There is a third quadrant spectral sequence converging to the homology of the above double complex with $E_2^{p,q} = H_{-p,-q}$ (row homology). If FP is F^i-acyclic (that is, $L_jF^i(FP) = 0$ for $j > 0$) for all $i \geq 1$ and all projectives P in $\underline{\underline{A}}$, then it is easily seen that $E_2^{-p,-q} = L_qF^p(A)$, for $p \geq 1$, $q \geq 0$.

Elementary calculations of edge homomorphisms together with the preceding remarks yield the following result (See Cartan and Eilenberg [14]).

Theorem 4.6. *If $\alpha : FA \longrightarrow A$ represents an object in $\underline{\underline{A}} \ltimes F$, there are natural maps $\eta_i : L_iC(\alpha) \longrightarrow H_i^F(\alpha)$ for $i \geq 0$ having the following properties:*

1) *η_i is an isomorphism for* $i = 0,1$.

2) *η_2 is an epimorphism.*

3) *If FP is F^r-acyclic* (See 4.5. (3d) above) *for all $r \geq 1$ and all projectives P in $\underline{\underline{A}}$ and if $L_qF^p = 0$ for $p+q \geq n+1$ $(p,q \geq 1)$, then $\eta_{n+i} : L_{n+i}C(\alpha) \longrightarrow H_{n+i}^F(\alpha)$ is an isomorphism for $i \geq 1$.*

4) *In particular, if F is exact, then η_i is an isomorphism for all $i \geq 0$.* QED.

Corollary 4.7. *Let $\underline{\underline{A}}$ be an abelian category with enough projectives and let $F : \underline{\underline{A}} \longrightarrow \underline{\underline{A}}$ be a right exact functor such that FP is F^r-acyclic for all projective P and all $r \geq 1$.*

a) *If $L_qF^p = 0$ for $p+q \geq n+1$ $(p \geq 1,\ q \geq 0)$, then $L_{n+1}C = 0$, where $C: \underline{\underline{A}} \ltimes F \longrightarrow \underline{\underline{A}}$ denotes the cokernel functor. In particular, if F is exact, then $L_nC = 0$ if and only if $F^n = 0$ $(n \geq 1)$.*

b) *If gl. dim $\underline{\underline{A}} = m < \infty$ and if $F^n = 0$, then $L_{n+m}C = 0$.*

Proof. Part a) is an obvious consequence of Theorem 4.6.(3), (4). And Part b) follows directly from part a). QED.

Having determined some conditions under which the left derived functors L_nC of the cokernel functor $C : \underline{\underline{A}} \ltimes F \longrightarrow \underline{\underline{A}}$ vanish for large n, we shall now make more explicit the connection between this phenomenon and the estimation of projective dimension of objects in $\underline{\underline{A}} \ltimes F$.

<u>Lemma</u> 4.8. <u>The object</u> $\alpha : FA \longrightarrow A$ <u>in</u> $\underline{\underline{A}} \ltimes F$ <u>is projective if and only if</u> $L_1C(\alpha) \cong H_1{}^F(\alpha) = 0$ <u>and</u> $\operatorname{cok} \alpha = C(\alpha)$ <u>is projective in</u> $\underline{\underline{A}}$.

Proof. The necessity is clear from the structure of projective objects in $\underline{\underline{A}} \ltimes F$ (See Section 1).

Suppose $L_1C(\alpha) \cong H_1{}^F(\alpha) = 0$ and $\operatorname{cok} \alpha$ is projective in $\underline{\underline{A}}$. It follows that the sequence $0 \longrightarrow \text{image } \alpha \longrightarrow A \longrightarrow \operatorname{cok} \alpha \longrightarrow 0$ is split exact and that the commutative triangle

$$\begin{array}{ccc} FA & \searrow & \\ \alpha \downarrow & & F(\operatorname{cok} \alpha) \\ A & \swarrow & \end{array}$$

in Observation 4.5(2), gives an isomorphism $F(\operatorname{cok} \alpha) \longrightarrow \text{image } \alpha$. Hence, $(FA \xrightarrow{\alpha} A) \underset{\underline{\underline{A}} \ltimes F}{\cong} T(\operatorname{cok} \alpha)$ and α represents a projective object in $\underline{\underline{A}} \ltimes F$. QED.

Our next theorem is a natural extension of Lemma 4.8. to objects in $\underline{\underline{A}} \ltimes F$ of finite projective dimension.

<u>Theorem</u> 4.9. <u>If</u> $\alpha : FA \longrightarrow A$ <u>represents an object in</u> $\underline{\underline{A}} \ltimes F$ <u>of finite projective dimension and if</u> $L_nC(\alpha) = 0$ <u>for all</u> $n \geq 1$, <u>then</u>

a) $L_nF(\operatorname{cok} \alpha) = 0$ <u>for all</u> $n \geq 1$,

b) $\operatorname{pd}_{\underline{\underline{A}} \ltimes F}(\alpha) = \operatorname{pd}_{\underline{\underline{A}}}(\operatorname{cok} \underline{\alpha})$.

Proof. If $\operatorname{pd}_{\underline{\underline{A}} \ltimes F}(\alpha) = 0$, that is α represents a projective object in $\underline{\underline{A}} \ltimes F$, then statements a) and b) above easily follow from Lemma 4.8. We shall continue by way of induction on $\operatorname{pd}_{\underline{\underline{A}} \ltimes F}(\alpha) = n \geq 1$. A one-step projective resolution of α gives the following commutative diagram

$$\begin{array}{ccccccccc} & & FB & \longrightarrow & FP \oplus F^2P & \longrightarrow & FA & \longrightarrow & 0 \\ & & \beta \downarrow & & \searrow = & & \downarrow \alpha & & \\ 0 & \longrightarrow & B & \longrightarrow & P \oplus FP & \longrightarrow & A & \longrightarrow & 0 \end{array}$$

where, of course, P is a projective object in $\underline{\underline{A}}$. Since $L_nC(\alpha) = 0$ for $n \geq 1$, it follows that

i) $0 \longrightarrow \operatorname{cok} \beta \longrightarrow P \longrightarrow \operatorname{cok} \alpha \longrightarrow 0$

is exact and,

ii) $L_nC(\beta = 0$ for all $n \geq 1$.

By induction, $pd_{\underline{\underline{A}} \ltimes F}(\alpha) = 1 + pd_{\underline{\underline{A}} \ltimes F}(\beta) = 1 + pd_{\underline{\underline{A}}}(\text{cok}\ \beta)$. Hence if $pd_{\underline{\underline{A}}}(\text{cok}\ \alpha) > 0$, then $pd_{\underline{\underline{A}}}(\text{cok}\ \alpha) = 1 + pd_{\underline{\underline{A}}}(\text{cok}\ \beta) = pd_{\underline{\underline{A}} \ltimes F}(\alpha)$. However $pd_{\underline{\underline{A}}}(\text{cok}\ \alpha) > 0$ is guaranteed by Lemma 4.8. and the fact that $pd_{\underline{\underline{A}} \ltimes F}(\alpha) > 0$.

It remains to establish Part a). Again by induction, we have (with reference to the above setting) that $L_iF(\text{cok}\ \beta) = 0$ for all $i > 0$. Moreover, since $H_1^F(\beta) \cong L_1C(\beta) = 0$, there is a commutative diagram (See Observation 4.5 (2))

$$\begin{array}{ccccccccc} & & F(\text{cok}\ \beta) & \longrightarrow & FP & \longrightarrow & F(\text{cok}\ \alpha) & \longrightarrow & 0 \\ & & \downarrow & & \searrow^{=} & & \downarrow & & \\ 0 & \longrightarrow & B & \longrightarrow & P \oplus FP & \longrightarrow & A & \longrightarrow & 0 \end{array}$$

where $F(\text{cok}\ \beta) \longrightarrow B$ is monic. Thus, we conclude that $F(\text{cok}\ \beta) \longrightarrow FP$ is monic and that $L_iF(\text{cok}\ \alpha) = 0$ for all $i > 0$. QED.

In subsequent parts of this section, we shall observe several situations where the following corollary holds.

Corollary 4.10. *Let* $\underline{\underline{A}}$ *be an abelian category with enough projectives, let* $F: \underline{\underline{A}} \longrightarrow \underline{\underline{A}}$ *be a right exact functor and let* $C: \underline{\underline{A}} \ltimes F \longrightarrow \underline{\underline{A}}$ *be the cokernel functor. If* L_iC *vanishes on all objects in* $\underline{\underline{A}} \ltimes F$ *of finite projective dimension, for all* $i \geq 1$, *then*

$$FPD(\underline{\underline{A}} \ltimes F) = \sup \{pd_{\underline{\underline{A}}}A < \infty : A \text{ is } F\text{-acyclic}\}.$$

In particular, $FPD(\underline{\underline{A}} \ltimes F) \leq FPD(\underline{\underline{A}})$.

Proof. The above statement is an immediate consequence of Theorem 4.9. and Lemma 4.4. QED.

Example 4.30. of Part C of this section shows that the inequality $FPD(\underline{\underline{A}} \ltimes F) \leq FPD(\underline{\underline{A}})$ may be strict.

For later reference, we record the dual statement of Theorem 4.9. for the category $G \rtimes \underline{\underline{A}}$ (G left exact).

Theorem 4.11. *If* $\beta : A \longrightarrow GA$ *represents an object in* $G \rtimes \underline{\underline{A}}$ *of finite injective dimension and if* $R^nK(\beta) = 0$ *for* $n \geq 1$ $(K : G \rtimes \underline{\underline{A}} \longrightarrow \underline{\underline{A}}$ is the kernel functor), *then*

a) $R^nG(\ker \beta) = 0$ *for all* $n \geq 1$,

and b) $id_{G \rtimes \underline{\underline{A}}}(\beta) = id_{\underline{\underline{A}}}(\ker \beta)$. QED.

Our main result on $FPD(\underline{\underline{A}} \ltimes F)$ and gl. $\dim(\underline{\underline{A}} \ltimes F)$ now follow.

Theorem 4.12. *Let* $\underline{\underline{A}}$ *be an abelian category with enough projectives and let* $\underline{\underline{A}} \ltimes F$ *be the trivial extension of* $\underline{\underline{A}}$ *by a right exact functor* $F: \underline{\underline{A}} \longrightarrow \underline{\underline{A}}$. *If the cokernel functor* $C: \underline{\underline{A}} \ltimes F \longrightarrow \underline{\underline{A}}$ *has the property that* $L_{n+1}C \equiv 0$, *then the following inequalities hold:*

1) $FPD(\underline{\underline{A}} \ltimes F) \leq n + FPD(\underline{\underline{A}})$.

2) $FPD(\underline{\underline{A}} \ltimes F) \geq \sup\{pd_{\underline{\underline{A}}}A < \infty : A$ is F-acyclic$\}$.

3) gl. $\dim \underline{\underline{A}} \leq$ gl. $\dim(A \ltimes F) \leq n +$ gl. $\dim \underline{\underline{A}}$.

Proof. First, we note that statement 2) is a consequence of Lemma 4.4. and necessarily holds regardless of whether or not $L_{n+1}C = 0$. Second, the first inequality in 3) was established in Corollary 4.3. Clearly, it remains to establish part 1). To this end, suppose α represents an object in $\underline{\underline{A}} \ltimes F$ of finite projective dimension. The "usual" dimension shift gives an object in $\underline{\underline{A}} \ltimes F$ such that $pd_{\underline{\underline{A}} \ltimes F}(\alpha) = n + pd_{\underline{\underline{A}} \ltimes F}(\beta)$ and $L_{n+1}C(\alpha) \cong L_iC(\beta) = 0$, for all $i \geq 1$. By Theorem 4.9, $pd_{\underline{\underline{A}} \ltimes F}(\beta) = pd_{\underline{\underline{A}}}\text{cok}\, \beta$. Hence, $pd_{\underline{\underline{A}} \ltimes F}(\alpha) \leq n + FPD(\underline{\underline{A}})$.

QED.

Corollary 4.13. (Same notation as 4.12). *If* FP *is* F^r-*acylic for all projective objects* P *in* $\underline{\underline{A}}$ *and all* $r \geq 1$, *and if* $L_qF^p = 0$ *for* $q + p \geq n + 1$ $(q \geq 0, p \geq 1)$, *then*

$$FPD(\underline{\underline{A}} \ltimes F) \leq n + FPD(\underline{\underline{A}}).$$

In particular, if F *is exact and* $F^{n+1} \equiv 0$, *then*

$$FPD(\underline{\underline{A}} \ltimes F) \leq n + FPD(\underline{\underline{A}}).$$

The above statements follow immediately from Corollary 4.7 and Theorem 4.12 QED.

Remarks 1. In case $F: \underline{\underline{A}} \longrightarrow \underline{\underline{A}}$ is an exact functor we have the natural isomorphism (Theorem 4.6(4))

$$L_{n+1}C(\alpha) \cong H^F_{n+1}(\alpha) \cong F^n(H_1{}^F(\alpha)) \quad \text{for} \quad n \geq 0,$$

where F^0 = identity. Hence, if gl. dim $\underline{\underline{A}} < \infty$, then the object α has finite projective dimension in $\underline{\underline{A}} \ltimes F$ if and only if $F^n(H^F_1(\alpha)) = 0$ for some $n \geq 1$.

2. Once it is known that gl. dim($\underline{\underline{A}} \ltimes F$) is finite, then the Palmer-Roos results [56,57] generally (with some exceptions in Part B of this section) give a more precise calculation of gl. dim($\underline{\underline{A}} \ltimes F$). Nevertheless, the next result seems to provide a more simple criterion that gl. dim($\underline{\underline{A}} \ltimes F$) be finite than those obtained by Palmer and Roos [56,57] under similar hypotheses.

Theorem 4.14. *Assume the right exact functor* $F: \underline{\underline{A}} \longrightarrow \underline{\underline{A}}$ *has the property that* FP *is* F^r*-acyclic for all projective* P *and all* $r \geq 1$. *Then* gl. dim($\underline{\underline{A}} \ltimes F$) $< \infty$ *if and only if*

a) gl. dim $\underline{\underline{A}} = m < \infty$, *and*

b) $F^n = 0$ for some $n \geq 1$.

In case a) and b) *hold, then*

$$m \leq \text{gl. dim}(\underline{\underline{A}} \ltimes F) \leq n + 2m - 1.$$

Proof. If a) and b) hold, then $L_{n+m}C = 0$, by Corollary 4.7(b). By Theorem 4.12(3), $m \leq$ gl. dim($\underline{\underline{A}} \ltimes F$) $\leq (n+m-1) + m$.

Now suppose gl. dim($\underline{\underline{A}} \ltimes F$) $= d < \infty$. Then gl. dim $\underline{\underline{A}} = m \leq$ by Corollary 4.3. The hypothesis of this theorem yields the following fact: If X in $\underline{\underline{A}}$ is F^r-acyclic for all r, then F^rX is F-acyclic for all r. Let P be any projective in $\underline{\underline{A}}$. The preceding statement implies that F^rP is F-acyclic for all $r \geq 1$. Therefore, we may lift (via the functor T) a projective reolution (finite) of F^rP to a projective resolution of $T(F^rP)$. The kernels of the latter resolution are of the form $T(X)$, for $X \in \underline{\underline{A}}$. By Observation 4.5(1) and Theorem 4.6(1), we obtain that $L_iC(T(F^rP) = 0$, for $i,r, \geq 1$. Hence, we now have a canonical C-acyclic resolution of the object

$\begin{array}{c} FP \\ \downarrow 0 \\ P \end{array}$ of the form

$$\cdots \longrightarrow \begin{array}{c} F^3P \oplus F^4P \\ \searrow = \\ F^2P \oplus F^3P \end{array} \longrightarrow \begin{array}{c} F^2P \oplus F^3P \\ \searrow = \\ FP \oplus F^2P \end{array} \longrightarrow \begin{array}{c} FP \oplus F^2P \\ \searrow = \\ P \oplus FP \end{array} \longrightarrow \begin{array}{c} FP \\ \downarrow 0 \\ P \end{array} \longrightarrow 0$$

which is the complex

$$\longrightarrow T(F^2P) \longrightarrow T(FP) \longrightarrow T(P) \longrightarrow ZP \longrightarrow 0.$$

It easily follows from the above resolution, Observation 4.5(1) and Theorem 4.6(1) that $L_i C \left(\begin{array}{c} FP \\ \downarrow 0 \\ P \end{array} \right) \cong F^i P$ for $i \geq 1$. But $L_{d+1} C = 0$, for $d = \text{gl. dim}\,(\underline{\underline{A}} \ltimes F)$. Thus $F^{d+1}P = 0$ for all projective objects P in $\underline{\underline{A}}$. However, since F^{d+1} is necessarily a right exact functor, it follows that $F^{d+1} = 0$. QED.

In Part C of this section (Example 4.30), we shall show the assumption that $F: \underline{\underline{A}} \longrightarrow \underline{\underline{A}}$ takes projective objects to F^r-acyclic objects, $r \geq 1$, cannot simply be dropped in Theorem 4.14 Moreover, if one is specifically considering the category of left $R \ltimes M$-modules, where M is an R- bimodule, then the above condition becomes

$$\text{Tor}_i^R(\underbrace{M \otimes_R M \otimes \ldots \otimes M}_{r}, M) = 0 \quad \text{for all} \quad i \geq 1 \quad \text{and} \quad r \geq 1.$$

(Here $F^r(X) = \underbrace{M \otimes_R M \otimes_R M \otimes_R \ldots \otimes_R M}_{r} \otimes_R X$.)

The following example demonstrates the instability of the $\text{FPD}(\underline{\underline{A}} \ltimes F)$ with respect to the $\text{FPD}(\underline{\underline{A}})$.

Example 4.15. Categories of Complexes. Let $\underline{\underline{A}}$ be an abelian category with enough projectives and let $\underline{\underline{A}}^N$ denote the countably infinite product category $\underline{\underline{A}} \times \underline{\underline{A}} \times \underline{\underline{A}} \times \ldots$. In this category an object is a sequence of objects in $\underline{\underline{A}}$,

$$A = (A_0, A_1, \ldots),$$

where $A_i \in \underline{\underline{A}}$, and morphisms act componentwise. In addition $\underline{\underline{A}}^N$ has enough projectives and clearly $FPD(\underline{\underline{A}}^N) = FPD(\underline{\underline{A}})$ and $gl. dim(\underline{\underline{A}}^N) = gl. dim(\underline{\underline{A}})$.

Let $F: \underline{\underline{A}}^N \longrightarrow \underline{\underline{A}}^N$ be the functor which takes the object $(A_0, A_1, A_2, \ldots)$ to the object $(A_1, A_2, \ldots)$ and which "shifts" morphisms in the same way. Clearly, F is an exact functor. (A sequence in $\underline{\underline{A}}$ is exact if and only if it is componentwise exact.) We let $C_+(\underline{\underline{A}})$ denote the trivial extension category $\underline{\underline{A}}^N \ltimes F$. If $d: FA \longrightarrow A$, $A = (A_0, A_1, \ldots)$, represents an object in $C_+(\underline{\underline{A}})$, the requirement $dF(d) = 0$, that is the composite map

$$\begin{array}{rl} F^2A & = (A_2, A_3, A_4, \ldots) \\ {\scriptstyle Fd}\downarrow & \\ F\,A & = (A_1, A_2, A_3, \ldots) \\ {\scriptstyle d}\downarrow & \\ A & = (A_0, A_1, A_2, \ldots) \end{array}$$

is zero, is equivalent to the requirement that A is a nonnegative complex over $\underline{\underline{A}}$ with differential d. Moreover, the remark following Corollary 4.13 gives the isomorphisms

4.15(a) $$L_{n+1}C(d) \cong H^F_{n+1}(d) \cong F^n(H^F_1(d)) = \coprod_{i \geq n+1} H_i(A,d),$$

for $n \geq 0$, where $H_i(A,d)$ denotes the $i^{\underline{th}}$ homology of the complex $A = (A_0, A_1, \ldots)$ with differential d.

Hence, if $gl. dim\, \underline{\underline{A}} < \infty$, then a complex (A,d) in $C_+(\underline{\underline{A}})$ has finite projective dimension if and only if $H_i(A,d) = 0$ for all $i \geq m$, for some $m \geq 0$ (depending on (A,d)). Moreover, $pd(A,d) \geq n$ if $H_n(A,d) \neq 0$. An easy consequence of these remarks is the following statement.

4.15(b). If $\underline{\underline{A}}$ is a nontrivial abelian category with enough projectives, then

$$FPD(C_+(\underline{\underline{A}})) = FPD(\underline{\underline{A}}^N \ltimes F) = \infty.$$

(even though, $gl. dim\, \underline{\underline{A}}^N = gl. dim\, \underline{\underline{A}} < \infty$).

Now let $\underline{\underline{A}} = \underline{\underline{Mod}}\ R$, where R is a (nontrivial) commutative ring. There is an exact full embedding of the category $(\underline{\underline{Mod}}\ R)^N$ into the category $\underline{\underline{Mod}}(R^N)$, where R^N is the countable infinite product of copies R. This embedding preserves projectives which are componentwise n-generated, for any $n \geq 1$. In addition, the functor $F(R^N) \otimes_R N$ (The object $F(R^N)$ is not symmetric!) agrees with the "shift" functor $F: (\underline{\underline{Mod}}\ R)^N \longrightarrow (\underline{\underline{Mod}}\ R)^N$ (defined above) on R^N-modules of the form $M = (M_0, M_1, \ldots)$, where each M_i is n-generated (n depends on M). $n \geq 1$. Thus, we have the following statement.

4.15 (c). Let R be a commutative ring and let $M = F(R^N)$ as defined above. Then $M \otimes_{R^N} -$ is an exact functor and left $FPD(R^N \ltimes M) = \infty$ (regardless of whether or not FPD(R) is finite or infinite). Moreover, if one assumes the Continuum Hypothesis and if R is a field, then $FPD(R^N) = \text{gl. dim}\ R^N = 1$ [See B. L. Osofsky; 54, 55].

Of course, we need not restrict our attention to the category of nonnegative complexes over an abelian category $\underline{\underline{A}}$ (with enough projectives). For example, the category of complexes $C_{n+1}(\underline{\underline{A}})$ of length $n+1$ over $\underline{\underline{A}}$, that is, of the form

$$A_0 \xleftarrow{d_1} A_1 \xleftarrow{d_2} A_2 \longleftarrow \cdots \longleftarrow A_{n-1} \longleftarrow A_n,$$

can be defined as a trivial extension of $\underline{\underline{A}}^{n+1}$ in a similar way. Previous considerations (above) as well as Theorem 4.12 yield our next calculation.

4.15 (d). $\quad FPD(C_{n+1}(\underline{\underline{A}})) = n + FPD(\underline{\underline{A}})$

and $\quad \text{gl. dim}\ C_{n+1}(\underline{\underline{A}}) = n + \text{gl. dim}\ (\underline{\underline{A}})$.

4.B. <u>The finitistic projective dimension for</u> $\underline{\underline{Map}}\ (F\underline{\underline{A}}, \underline{\underline{B}})$ <u>and triangular matrix rings</u>.

Let $\underline{\underline{A}}$ and $\underline{\underline{B}}$ be abelian categories and let $F: \underline{\underline{A}} \longrightarrow \underline{\underline{B}}$ be a right exact functor. Then $\underline{\underline{M}} = \underline{\underline{Map}}(F\underline{\underline{A}}, \underline{\underline{B}})$ denotes the category of maps $FA \xrightarrow{f} B$, where $A \in \underline{\underline{A}}$ and $B \in \underline{\underline{B}}$ (See Section 1 for details). As it has already been noted in Section 1, $\underline{\underline{M}}$ is isomorphic to the category $(\underline{\underline{A}} \times \underline{\underline{B}}) \ltimes \tilde{F}$, where $\underline{\underline{A}} \times \underline{\underline{B}}$ is the product category and where the functor $\tilde{F}: \underline{\underline{A}} \times \underline{\underline{B}} \longrightarrow \underline{\underline{A}} \times \underline{\underline{B}}$ is defined as follows:

i) $\tilde{F}(A,B) = (0,FA)$

ii) If $(A,B) \xrightarrow{(\alpha,\beta)} (A', B')$ is a morphism in $\underline{\underline{A}} \times \underline{\underline{B}}$, then $\tilde{F}(\alpha,\beta) = (0, F\alpha)$.

We briefly review some of the properties of F needed for calculation of projective dimension in $\underline{\underline{M}} = \underline{\underline{Map}}(F\underline{\underline{A}}, \underline{\underline{B}})$.

i) $\tilde{F}^2 = 0$.

ii) $L_i\tilde{F}(A,B) \cong (0, L_iFA)$ (natural isomorphism).

In particular, $L_i\tilde{F}$ vanishes on all objects of the form

$$\tilde{F}(A,B) = (0,FA).$$

iii) $\tilde{F}$ is exact on short exact sequences of the form

$$0 \longrightarrow (0,B') \longrightarrow (0,B) \longrightarrow (0,B'') \longrightarrow 0.$$

We observe that properties i) and ii) are reminiscent of those in Part A of this section which were needed in the calculation of the FPD of trivial extension categories.

Lemma 4.16. Let $\alpha : \tilde{F}(A,B) \longrightarrow (A,B)$ represent an object in $\underline{\underline{M}} = \underline{\underline{Map}}\ (F\underline{\underline{A}}, \underline{\underline{B}})$. Then the following inequalities hold:

a) $pd_{\underline{\underline{M}}}(\alpha) \geq pd_{\underline{\underline{A}}}A$.

b) If $A = 0$, then $pd_{\underline{\underline{M}}}(\alpha) = pd_{\underline{\underline{B}}}B$.

c) Assume $L_iF(A) = 0$ for all $i > 0$. If $B = FA$ and $\alpha = (0, id_{FA})$, then $pd_{\underline{\underline{M}}}(\alpha) = pd_{\underline{\underline{A}}}(A)$.

Proof. In any case, we note that the image of α is contained in $(0,B)$. Hence, part a) is a direct consequence of Lemma 4.1.

As usual, let $T : \underline{\underline{A}} \times \underline{\underline{B}} \longrightarrow \underline{\underline{M}}$ be the tensor functor as defined in Section 1, that is,

$$T(A,B) = \begin{array}{c} \tilde{F}(A,B) \oplus \tilde{F}^2(A,B) \\ \searrow \quad = \\ (A,B) \oplus \tilde{F}(A,B). \end{array}$$

From the properties of $\tilde{F}$ listed above, we see that

$$T(0,B) = \begin{array}{c}(0,0)\\ \downarrow 0\\ (0,B)\end{array} \quad \text{and} \quad T(A,0) = \begin{array}{c}(0,FA)\\ \downarrow (0,id_{FA})\\ (A,FA)\end{array}$$

Hence, parts b) and c) are immediate consequences of Lemma 4.4. QED.

As in Part A of this section, we let $C\colon \underline{\underline{M}} = (\underline{\underline{A}} \times \underline{\underline{B}}) \ltimes \tilde{F} \longrightarrow \underline{\underline{A}} \times \underline{\underline{B}}$ denote the cokernel functor. In the next two lemmas, we derive several properties of C (from Part A) in our present setting.

Lemma 4.17. *Let $\alpha : (0,FA) \longrightarrow (A,B)$ represent an object in* $\underline{\underline{M}} = \underline{\underline{\text{Map}}}\ (F\underline{\underline{A}}, \underline{\underline{B}})$.

a) $L_1C(\alpha) \cong \text{Ker}\ \alpha \subseteq (0,FA)$;

hence $L_iC(\alpha) \subseteq 0 \times \underline{\underline{B}}$, *for all* $i > 0$.

b) L_iC *vanishes on objects of the form* $(0,0) \longrightarrow (0,B)$, *for* $i > 0$.

c) *The exact sequence*

$$\begin{array}{ccccccccc} & & (0,0) & \longrightarrow & (0,FA) & \longrightarrow & (0,FA) & \longrightarrow & 0\\ & & \downarrow & & \downarrow \alpha & & \downarrow 0_A & & \\ 0 & \longrightarrow & (0,B) & \longrightarrow & (A,B) & \longrightarrow & (A,0) & \longrightarrow & 0 \end{array}$$

induces a map $L_iC(\alpha) \longrightarrow L_iC(0_A)$ *which is an isomorphism for* $i \geq 2$ *and a monomorphism for* $i = 1$.

d) *If* F *is exact, then* $L_iC = 0$ *for* $i \geq 2$.

Proof. Part a) of this lemma follows directly from Theorem 4.6(1). To see b), it suffices to observe that, if $Q_\bullet$ is a $\underline{\underline{B}}$-projective resolution of B, then $\begin{array}{c}(0,0)\\ \downarrow\\ (0,Q_\bullet)\end{array}$ is an $\underline{\underline{M}}$-projective resolution of $\begin{array}{c}(0,0)\\ \downarrow\\ (0,B)\end{array}$. Part c) is an immediate consequence of b). Finally, statement d) is a consequence of Theorem 4.6(4) and the fact that $\tilde{F}$ is exact if and only if F is exact. QED.

In the sequel, for $A \in \underline{\underline{A}}$, 0_A always denotes the object $(0,FA) \xrightarrow{0} (A,0)$ in $\underline{\underline{M}}$ and, for $B \in \underline{\underline{B}}$, 0_B denotes the object $(0,0) \to (0,B)$

in $\underline{\underline{M}}$. There is always an $\underline{\underline{M}}$-exact sequence

$$0 \longrightarrow 0_B \longrightarrow \alpha \longrightarrow 0_A \longrightarrow 0,$$

where $\alpha: (0,FA) \longrightarrow (A,B)$ is in $\underline{\underline{M}}$.

Lemma 4.18. Let $\alpha = (0,f) : (0,FA) \longrightarrow (A,B)$ represent an object in $\underline{\underline{M}} = \mathrm{Map}(F\underline{\underline{A}},\underline{\underline{B}})$.

1) If $L_iC(\alpha) = 0$ for all $i > 0$,
then $pd_{\underline{\underline{M}}}(\alpha) = \max\,(pd_{\underline{\underline{A}}}A,\ pd_{\underline{\underline{B}}}\mathrm{cok}\ f)$.

2) If $pd_{\underline{\underline{A}}}A \leq n$,
then $L_iC(0_A) = 0$ for $i \geq n + 2$.

Proof. Statement 1) is an immediate consequence of Theorem 4.9.

To see 2), suppose A is a projective object in $\underline{\underline{A}}$. There is an exact sequence in $\underline{\underline{M}}$

$$\begin{array}{ccccccccc} & & (0,0) & \longrightarrow & (0,FA) & \longrightarrow & (0,FA) & \longrightarrow & 0 \\ & & \downarrow & & \downarrow{\scriptstyle (0,id_{FA})} & & \downarrow{\scriptstyle 0_A} & & \\ 0 & \longrightarrow & (0,FA) & \longrightarrow & (A,FA) & \longrightarrow & (A,0) & \longrightarrow & 0 \end{array}$$

where of course, the middle term is $\underline{\underline{M}}$-projective. From Lemma 4.17 c), we now see that $L_iC(0_A) = 0$ for $i \geq 2$. The remainder of the proof follows by the "usual" induction on $pd_{\underline{\underline{A}}}A$. QED.

A large portion of what follows (in part B) is a result of the next theorem.

Theorem 4.19. If $\alpha: (0,FA) \longrightarrow (A,B)$ represents an object in $\underline{\underline{M}} = \mathrm{Map}(F\underline{\underline{A}},\ \underline{\underline{B}})$ of finite projective dimension, then $pd_{\underline{\underline{M}}}\alpha \geq pd_{\underline{\underline{A}}}A$ and there is an object B' in $\underline{\underline{B}}$ such that

$$pd_{\underline{\underline{A}}}A \leq pd_{\underline{\underline{M}}}(\alpha) \leq 1 + pd_{\underline{\underline{A}}}A + pd_{\underline{\underline{B}}}B' < \infty .$$

Proof. That $pd_{\underline{\underline{M}}}(\alpha) \geq pd_{\underline{\underline{A}}}A$ is a restatement of Lemma 4.16(a). So let $pd_{\underline{\underline{A}}}A = n$ (finite). From Lemma 4.18(2) and Lemma 4.17(c), we see that $L_iC(\alpha) = 0$ for $i \geq n + 2$. Next we observe that

if
$$\begin{array}{c}(0,FP_\bullet)\\ \downarrow f_\bullet \\ (P_\bullet, X_\bullet)\end{array}$$

is an $\underline{\underline{M}}$-projective resolution of $\begin{array}{c}(0,FA)\\ \downarrow \alpha \\ (A,B)\end{array}$, then $P_\bullet$ is a projective resolution of A in $\underline{\underline{A}}$. Now let $\beta = (0,g) : (0,FK) \to (K,N)$ be the $(n+1)^{\underline{st}}$ syzygy in an $\underline{\underline{M}}$-projective resolution of α. From above $L_iC(\beta) = 0$ for all $i > 0$. Hence, $pd_{\underline{\underline{M}}}\beta = \max(pd_{\underline{\underline{A}}}K, pd_{\underline{\underline{B}}} \operatorname{cok} g)$ by Lemma 4.18 (1). However, our preceding remark on $\underline{\underline{M}}$-projective resolutions, ensures that $pd_{\underline{\underline{A}}}K = 0$. Therefore, $pd_{\underline{\underline{M}}}\beta = pd_{\underline{\underline{B}}}(\operatorname{cok} g)$. Setting $B' = \operatorname{cok} g$, we obtain our desired conclusion. QED.

Theorem 4.20. Let $\underline{\underline{M}} = \underline{\underline{Map}}(F\underline{\underline{A}},\underline{\underline{B}})$. Then the following inequalities hold:

1) $FPD(\underline{\underline{B}}) \leq FPD(\underline{\underline{M}}) \leq 1 + FPD(\underline{\underline{A}}) + FPD(\underline{\underline{B}})$.

2) $FPD(\underline{\underline{M}}) \geq \sup\{pd_{\underline{\underline{A}}}A < \infty : L_iF(A) = 0 \text{ for } i > 0\}$

If F is exact, then $FPD(\underline{\underline{M}}) \geq FPD(\underline{\underline{A}})$.

3) $\max(\text{gl. dim } \underline{\underline{A}}, \text{gl. dim } \underline{\underline{B}}) \leq \text{gl. dim } \underline{\underline{M}} \leq 1 + \text{gl. dim } \underline{\underline{A}} + \text{gl. dim } \underline{\underline{B}}$.

Proof. Statement 1) is an immediate consequence of Lemma 4.16(b) and Theorem 4.19. In order to see 2), we observe that, if $L_iF(A) = 0$ for $i > 0$, then $L_i\tilde{F}(A,0) = 0$ for $i > 0$. An application of Lemma 4.4 now gives the desired inequality in 2).

Since $\text{gl. dim } (\underline{\underline{A}} \times \underline{\underline{B}}) = \max(\text{gl. dim } \underline{\underline{A}}, \text{gl. dim } \underline{\underline{B}})$, the left-hand inequality in 3) follows from Corollary 4.3. The right-hand inequality in 3) is a consequence of part 1). QED.

Remark. Example 4.27 at the end of this section (Part B) shows that, in general, $FPD(\underline{\underline{M}}) \not\geq FPD(\underline{\underline{A}})$, where $\underline{\underline{M}} = \underline{\underline{Map}}(F\underline{\underline{A}},\underline{\underline{B}})$. Hence, the claim $FPD(\underline{\underline{M}}) \geq FPD(\underline{\underline{A}})$ in Corollary III (a) of our previous paper [2] is in error.

We shall now interpret our results (with a few improvements in the calculation of global dimension) for triangular matrix rings.

Corollary 4.21. *Let* R *and* S *be rings, let* $M \neq 0$ *be an* S-R *bimodule and let* $\Lambda = \begin{pmatrix} R & 0 \\ M & S \end{pmatrix}$. *Then the following inequalities hold*:

1) left $\mathrm{FPD}(S) \leq$ left $\mathrm{FPD}(\Lambda) \leq 1 +$ left $\mathrm{FPD}(R)$ + left $\mathrm{FPD}(S)$.

2) left $\mathrm{FPD}(\Lambda) \geq \sup \{\mathrm{pd}_R A < \infty :$ A is a left R-module satisfying $\mathrm{Tor}_i^R(M,A) = 0$, for $i > 0\}$. *If* M *is flat as a right* R-*module, then* left $\mathrm{FPD}(\Lambda) \geq$ left $\mathrm{FPD}(R)$.

3) *If* $\mathrm{pd}_S M < \infty$, *then* $\mathrm{pd}_S M + 1 \leq$ left $\mathrm{FPD}(\Lambda) \leq$ max (left $\mathrm{FPD}(R) + \mathrm{pd}_S M + 1$, left $\mathrm{FPD}(S)$).

4) max (left gl. dim R, left gl. dim S, $\mathrm{pd}_S M + 1$) $\leq$ left gl. dim $\Lambda \leq$ max(left gl. dim R + $\mathrm{pd}_S M + 1$, left gl. dim S

Corresponding statements hold for the right homological dimensions over Λ.

Proof. Statements 1) and 2) are merely a rephrasing of Theorem 4.20(1), (2) in the context of triangular matrices.

We shall now make use of the isomorphism of categories ${}_\Lambda\mathrm{Mod} \cong \mathrm{Map}({}_F{}_R\mathrm{Mod}, {}_S\mathrm{Mod})$, where $F = M \otimes_R -$. The $\mathrm{Map}({}_F{}_R\mathrm{Mod}, {}_S\mathrm{Mod}) \cong {}_\Lambda\mathrm{Mod}$ exact sequence

$$\begin{array}{ccccccccc} & & (0,0) & \longrightarrow & (0,M) & \longrightarrow & (R,M) & \longrightarrow & 0 \\ & & \downarrow 0_M & & \downarrow (0,\mathrm{id}_M) & & \downarrow 0_R & & \\ 0 & \longrightarrow & (0,M) & \longrightarrow & (R,M) & \longrightarrow & (R,0) & \longrightarrow & 0 \end{array}$$

provides a Λ-projective cover of 0_R (See Corollary 1.7). Hence since $M \neq P$, the above Λ-exact sequence cannot split, and so $\mathrm{pd}(0_R) = 1 + \mathrm{pd}(0_M) = 1 + \mathrm{pd}_S M$. In the same spirit as the preceding argument, it is easy to verify (via induction) that

$$\mathrm{pd}\ 0_A \leq \mathrm{pd}_S M + 1 + \mathrm{pd}_R A, \quad \text{for } A \in {}_R\mathrm{Mod}.$$

Thus, statements 3) and 4) now follow. QED.

Remark. Palmer and Roos [56] provide more elaborate and precise inequalties than in statement 4) of Corollary 4.21.

We also remark that the finiteness of (left) FPD(Λ), $\Lambda = \begin{pmatrix} R & 0 \\ M & S \end{pmatrix}$, depends only on the finiteness of (left) FPD(R) and (left) FPD(S) and not on any peculiar properties of M.

Before considering several examples, we shall refine a few of the estimates in Theorem 4.20 and Corollary 4.21 in the case $F : \underline{\underline{A}} \to \underline{\underline{B}}$ is an exact functor.

Proposition 4.21. Let $\underline{\underline{M}} = \underline{\text{Map}}\,(F\underline{\underline{A}}, \underline{\underline{B}})$ and assume F is an exact functor. Then for $A \in \underline{\underline{A}}$,

$$pd_{\underline{\underline{M}}}(0_A) = \max(1 + pd_{\underline{\underline{B}}}(FA),\ pd_{\underline{\underline{A}}}A).$$

Proof. We adhere to the convention that the projective dimension of the zero object is $-\infty$ and that $1 + (-\infty) = -\infty$. Hence, it is clear that the above conclusion holds when A is projective in $\underline{\underline{A}}$. Moreover, Lemma 4.16(a) shows

$$pd_{\underline{\underline{M}}}(0_A) = \max\,(1 + pd_{\underline{\underline{B}}}(FA),\ pd_{\underline{\underline{A}}}A),$$

whenever $pd_{\underline{\underline{A}}}A = \infty$. Thus, in order to complete the proof, we may assume that $pd_{\underline{\underline{A}}}A$ is a positive integer n. Let

$$0 \to K \xrightarrow{i} P \to A \to 0$$

be exact in $\underline{\underline{A}}$, where P is A-projective. Then we obtain the $\underline{\underline{M}}$-exact sequence

$$\begin{array}{ccccccccc}
 & & (0,FK) & \to & (0,FP) & \to & (0,FA) & \to & 0 \\
 & & \downarrow{\scriptstyle (0,Fi)=\beta} & & \downarrow{\scriptstyle (0,id_{FP})} & & \downarrow{\scriptstyle 0_A} & & \\
0 & \to & (K,FP) & \to & (P,FP) & \to & (A,0) & \to & 0
\end{array}$$

It follows that $pd_{\underline{\underline{M}}}(0_A) = 1 + pd_{\underline{\underline{M}}}(\beta)$. Since $L_iC = 0$, $i \geq 2$, where $C\colon \underline{\underline{M}} \to \underline{\underline{A}} \times \underline{\underline{B}}$ is the cokernel functor (see Lemma 4.17(d)), we have that $L_iC(\beta) = 0$, for all $i > 0$. By Lemma 4.18(1), $pd_{\underline{\underline{M}}}(\beta) = \max\,(pd_{\underline{\underline{B}}}\text{cok}(Fi),\ pd_{\underline{\underline{A}}}K)$.

Thus,
$$pd_{\underline{\underline{M}}}(0\ ) = 1 + \max\,(pd_{\underline{\underline{B}}}\text{cok}(Fi),\ pd_{\underline{\underline{A}}}K)$$

$$= 1 + \max(pd_{\underline{\underline{B}}}(FA),\ pd_{\underline{\underline{A}}}K)$$

$$= \max(1 + pd_{\underline{\underline{B}}}(FA),\ 1 + pd_{\underline{\underline{A}}}K)$$

$$= \max(1+ pd_{\underline{\underline{B}}}(FA),\ pd_{\underline{\underline{A}}}A). \qquad \text{QED.}$$

If $\alpha: (0,FA) \longrightarrow (A,B)$ is an object in $\underline{\underline{M}} = \underline{\underline{Map}}(F\underline{\underline{A}},B)$ we always have the exact sequence

$$\begin{array}{ccccccccc} & & (0,0) & \longrightarrow & (0,FA) & \longrightarrow & (0,FA) & \longrightarrow & 0 \\ & & \downarrow 0_B & & \downarrow \alpha & & \downarrow \beta & & \\ 0 & \longrightarrow & (0,B) & \longrightarrow & (A,B) & \longrightarrow & (A,0) & \longrightarrow & 0. \end{array}$$

Thus, in case $F : \underline{\underline{A}} \longrightarrow \underline{\underline{B}}$ is exact, Proposition 4.21 and Lemma 4.16(b) allow us to determine rather precisely the various homological dimension in $\underline{\underline{M}} = \underline{\underline{Map}}(F\underline{\underline{A}},\underline{\underline{B}})$.

Corollary 4.22. _Let_ $\underline{\underline{M}} = \underline{\underline{Map}}(FA,B)$ _with_ $F : \underline{\underline{A}} \longrightarrow \underline{\underline{B}}$ _an exact functor_.

1) _If_ $FPD(\underline{\underline{B}}) = \infty$, _then_ $FPD(\underline{\underline{M}}) = \infty$.

2) _Assume_ $FPD(\underline{\underline{B}}) < \infty$. _If_ $pd_{\underline{\underline{B}}}FA < FPD(\underline{\underline{B}})$ _whenever_ $pd_{\underline{\underline{M}}}(0_A) < \infty$, _then_

$$FPD(\underline{\underline{M}}) = \max\ (FPD(\underline{\underline{B}}),\ FPD(\underline{\underline{A}})); \quad \textit{otherwise}$$

$$FPD(\underline{\underline{M}}) = \max\ (1 + FPD(\underline{\underline{B}}),\ FPD(\underline{\underline{A}})).$$

The above statements hold when "FPD" _is replaced by_ "gl. dim". QED.

Remark 4.23. Let $\underline{\underline{M}} = \underline{\underline{Map}}(F\underline{\underline{A}},\underline{\underline{B}})$. In general, the computation of the functor $Ext^1_{\underline{\underline{M}}}(-,-)$ requires the use of spectral sequences (See Palmer and Roos, [56,57]). However, if $F : \underline{\underline{A}} \longrightarrow \underline{\underline{B}}$ is an exact functor more elementary techniques may be used. We shall give a brief account here.

Recall the tensor functor $T : \underline{\underline{A}} \times \underline{\underline{B}} \longrightarrow \underline{\underline{M}}$ (See Section 1), wher

$$T(A,B) = \begin{array}{c}(0,FA)\\ \downarrow (0,i_{FA})\\ (A,FA \oplus B)\end{array}$$

and where i_{FA} is the natural injection of FA into $FA \oplus B$. We also recall that T preserves projective objects and that T is exact if and only if F is exact. Let $\beta : (0,FA') \longrightarrow (A',B')$ represent an object in $\underline{\underline{M}}$. There is a natural isomorphism

$$\mathrm{Hom}_{\underline{\underline{M}}}(T(A,B),\beta) \cong \mathrm{Hom}_{\underline{\underline{A}}}(A,A') \times \mathrm{Hom}_{\underline{\underline{B}}}(B,B').$$

Hence, if F is exact, the preceding natural isomorphism gives ...

4.23(a) $\mathrm{Ext}^i_{\underline{\underline{M}}}(T(A,B),\beta) \cong \mathrm{Ext}^i_{\underline{\underline{A}}}(A,A') \times \mathrm{Ext}^i_{\underline{\underline{B}}}(B,B')$

for all $i \geq 0$.

If $\alpha = (0,f) : (0,FA) \longrightarrow (A,B)$ is monic, there is an $\underline{\underline{M}}$-exact sequence

$$0 \longrightarrow T(A,0) \longrightarrow \alpha \longrightarrow T(0,\mathrm{cok}\ f) \longrightarrow 0.$$

This exact sequence together with 4.23(a) yields our next assertion.

4.23(b). If F is exact and if $\alpha = (0,f) : (0,FA) \to (A,B)$ is monic, then there is an exact sequence

$$\longrightarrow \mathrm{Ext}^i_{\underline{\underline{B}}}(\mathrm{cok}\ f,B') \longrightarrow \mathrm{Ext}^i_{\underline{\underline{M}}}(\alpha,\beta) \longrightarrow \mathrm{Ext}^i_{\underline{\underline{A}}}(A,A')$$

$$\longrightarrow \mathrm{Ext}^{i+1}_{\underline{\underline{B}}}(\mathrm{cok}\ f,B') \longrightarrow \ldots, \quad \text{for } i \geq 0$$

and for all $\beta : (0,FA') \longrightarrow (A',B')$ in $\underline{\underline{M}}$.

Finally, the exact sequence

$$\begin{array}{ccccccccc}
 & & (0,0) & \longrightarrow & (0,FA) & \longrightarrow & (0,FA) & \longrightarrow & 0\\
 & & \downarrow & & \downarrow & & \downarrow & & \\
0 & \longrightarrow & (0,FA) & \longrightarrow & (A,FA) & \longrightarrow & (A,0) & \longrightarrow & 0\\
 & & \parallel & & \parallel & & & & \\
0_{FA} & = & T(0,FA) & \rightarrow & T(A,0) & & & &
\end{array}$$

together with 4.23(a), (b) give the natural isomorphism ...

4.23(c). If F is exact, if $A \in \underline{\underline{A}}$ and if $B \in \underline{\underline{B}}$, then

$$\mathrm{Ext}^{i+1}_{\underline{\underline{M}}}(0_A, 0_B) \cong \mathrm{Ext}^{i}_{\underline{\underline{B}}}(FA, B),$$

for all $i \geq 0$. Furthermore, it can be shown that this isomorphism holds for $i = 0$ even if F is not exact.

We now end this part of section 4 with four examples which illustrate various aspects of the theory we have so far developed.

Example 4.24. Let R be any ring and let $T_n(R)$ denote the ring of $n \times n$ lower triangular matrices over R. If $n \geq 2$, then left $\mathrm{FPD}(T_n(R)) = 1 +$ left $\mathrm{FPD}(R)$. The same statement holds for left global dimension (and also for the right homological dimensions). The proofs of these equalities simply require Corollary 4.22 and induction. That left gl. dim $T_n(R) = 1 +$ left gl. dim (R) was first obtained by Eilenberg, Rosenberg and Zelinsky [17].

Example 4.25. (Difference of left and right global dimensions). In [35] Jategaonkar constructed rings R such that

$$(\text{right gl. dim } R) - (\text{left gl. dim } R) = n,$$

where n is a preassigned positive integer or ∞. We shall now give a rather simple construction of such rings in case n is finite.

Let S be a commutative Noetherian integral domain with field of quotients Q such that $2 \leq \mathrm{pd}_S Q = \text{gl. dim } S = n < \infty$ [See B. Osofsky Corollary 6.8; 54] and let $R = Q[X_1, \ldots, X_m]$ (ring of polynomials in m indeterminates) with $1 \leq m \leq n$. Since Q is an S-R bimodule in a natural way, we can form the ring $\Lambda = \begin{pmatrix} R & 0 \\ Q & S \end{pmatrix}$. Then left gl. dim $\Lambda =$ $= n + m + 1$ and right gl. dim $\Lambda = n$.

Since gl. dim $S = n > m =$ gl. dim R and since $-\otimes_S Q : \underline{\underline{\mathrm{Mod}}}\ S \longrightarrow \underline{\underline{\mathrm{Mod}}}\ R$ is an exact functor, it follows from Corollary 4.22(2) that right gl. dim $\Lambda = n$. Moreover, Corollary 4.21 gives left gl. dim $\Lambda \leq n + m + 1$. Hence, it remains to show left gl. dim $\Lambda \geq n + m + 1$. To accomplish this end, we resort to the isomorphism of categories ${}_{\Lambda}\underline{\underline{\mathrm{Mod}}} \cong \underline{\underline{\mathrm{Map}}}(F\,\underline{\underline{\mathrm{Mod}}}\,R, \underline{\underline{\mathrm{Mod}}}\ S)$, where $F = Q \otimes_R -$.

We claim, if A is an R-module of R-projective dimension $d \geq 1$ such that $L_iF(A) = \mathrm{Tor}^R_i(Q, A) = 0$ if and only if $i \geq d + 1$, then

$pd_\Lambda\ T(A,0) \geq n + d + 1$, where as usual

$$T(A,0) = \begin{array}{c}(0,FA)\\ \downarrow (0,id_{FA})\\ (A,FA)\end{array}.$$

Let A satisfy the above condition and let $0 \to K \to P \to A \to 0$ be exact in $\underline{\underline{Mod}}\ R$ with P an R-projective. Hence,

$$pd_R A = pd_R K + 1.$$

There is a $_\Lambda\underline{\underline{Mod}}$-exact sequence

$$\begin{array}{ccccccccc}
 & & (0,0) & \longrightarrow & (0,FK) & \longrightarrow & (0,FP) & \longrightarrow & (0,FA) \longrightarrow 0\\
 & & \downarrow & & \downarrow & & \downarrow & & \downarrow\\
0 & \longrightarrow & (0,L_1FA) & \longrightarrow & (K,FK) & \longrightarrow & (P,FP) & \longrightarrow & (A,FA) \longrightarrow 0\\
 & & \| & & \| & & \| & & \|\\
 & & T(0,L_1FA) & & T(K,0) & & T(P,0) & & T(A,0)
\end{array}$$

Suppose $pd_R A = 1$. By assumption, $L_1FA = Tor_1^R(Q,A) \neq 0$ and is necessarily isomorphic with a direct sum of copies of Q as an S-module. Hence,

$$n = pd_S(L_1FA) = pd_\Lambda\ T(0,L_1FA)$$

(See Lemma 4.16(6). When $pd_R A = 1$, the middle two terms $T(K,0)$ and $T(P,0)$ in the above exact sequence are Λ-projective. Therefore, $pd_\Lambda\ T(A,0) \geq n + 2$. An induction argument now gives the general case.

Let $A = R/X_1R \oplus R/(X_1,X_2)R \oplus \ldots \oplus R/(X_1, \ldots ,X_m)R$. Then $pd_R A = m$ and $Tor_i^R(Q,A) \neq 0$ if $m \geq i$. The discussion in the preceding paragraph gives $pd_\Lambda\ T(A,0) \geq n + m + 1$. Thus, left gl. dim $\Lambda \geq n + m + 1$.

Example 4.26. One can easily deduce from [M. Auslander; 1] the following statement: If P is a projective generator for $_R\underline{\underline{Mod}}$, R a ring, then left gl. dim $R = \sup\{pd_R X : X$ is of finite P type$\}$. A module is of finite P-type if and only if it is a homomorphic image of a finite direct sum of copies of P. In this example we show that, for R and S rings, $\underline{\underline{M}} = \underline{\underline{Map}}(F_R\underline{\underline{Mod}}, {}_S\underline{\underline{Mod}})$ need not inherit this property.

Let $\mathbb{Z}$ denote the ring of integers, let Q denote the field of rational numbers, let

$$F = \mathrm{Ext}^{\wedge}_{\mathbb{Z}}(\mathbb{Q},\_) : \mathrm{Mod}\ \mathbb{Z} \longrightarrow \mathrm{Mod}\ \mathbb{Q} \quad \text{and let}$$

$$\underline{\underline{M}} = \underline{\underline{\mathrm{Map}}}(F\ \underline{\underline{\mathrm{Mod}}}\ \mathbb{Z},\ \mathrm{Mod}\ \mathbb{Q}). \quad \text{Let}$$

$$P = T(\mathbb{Z},\mathbb{Q}) = \begin{array}{c}(0, F\mathbb{Z}) \\ \downarrow (0,\ \mathrm{incl}) \\ (\mathbb{Z},\ F\mathbb{Z} \oplus \mathbb{Q})\end{array}$$

It is easily checked (See Section 1) that P is a projective generator of $\underline{\underline{M}}$. The following (easily verified) statements concerning $\underline{\underline{M}}$ show that $\mathrm{gl.\,dim}\ \underline{\underline{M}} = 2$, while

$$\mathrm{Sup}\{\mathrm{pd}_{\underline{\underline{M}}} X : X \text{ is of finite P-type}\} = 1.$$

a) An object $\alpha = (0,F) : (0,FA) \longrightarrow (A,B)$

in $\underline{\underline{M}}$ is of finite P-type if and only if A is a finitely generated $\mathbb{Z}$-module and $\mathrm{cok}\ f$ is a finite generated $\mathbb{Q}$-module. Moreover, α is the direct sum of

$$(0,f') : (0,FA) \longrightarrow (A,B') \text{ and } T(0,\ \mathrm{cok}\ f),$$

where $B' = \mathrm{image}\ f$ and f' is the epimorphism induced by f.

b) $\mathrm{pd}_{\underline{\underline{M}}} T(0,B) = 0$, for $\mathbb{Q}$-modules B.

c) Since $L_1F \cong \mathrm{Hom}_{\mathbb{Z}}(\mathbb{Q},-)$, then

F is exact on finitely generated $\mathbb{Z}$-modules. Hence, $\mathrm{pd}_{\underline{\underline{M}}} T(A,0) = \mathrm{pd}_{\mathbb{Z}} A \leq 1$ (See Lemma 4.16(c)) whenever A is a finitely generated $\mathbb{Z}$-module.

d) If $\alpha = (0,f) : (0,FA) \longrightarrow (A,B)$

represents an object of finite P-type in $\underline{\underline{M}}$ with f an epimorphism, then the $\underline{\underline{M}}$ exact sequence

$$\begin{array}{ccccccccc} & & (0,0) & \longrightarrow & (0,FA) & \longrightarrow & (0,FA) & \longrightarrow & 0 \\ & & \downarrow & & \downarrow & & \downarrow \alpha & & \\ 0 & \longrightarrow & (0,\ \ker f) & \longrightarrow & (A,FA) & \longrightarrow & (A,B) & \longrightarrow & 0 \\ & & \| & & \| & & & & \\ & & T(0,\ker f) & & T(A,0) & & & & \end{array}$$

together with b) and c) show that $\mathrm{pd}_{\underline{\underline{M}}}(\alpha) \leq 1$.

Thus a), b), c) and d) yield

$$\sup\{pd_{\underline{\underline{M}}}X : X \text{ is of finite P-type}\} = 1$$

e) Let $0 \to P_1 \to P_o \to Q \to 0$ be a free $\mathbb{Z}$-resolution of $\mathbb{Q}$. An analysis (similar to that in Example 4.25) of the $\underline{\underline{M}}$-exact sequence

$$\begin{array}{ccccccccc} & & (0,0) & \to & (0,FP_1) & \to & (0,FP_o) & \to & (0,FQ) & \to 0 \\ & & \downarrow & & \downarrow & & \downarrow & & \downarrow & \\ 0 & \to & (0,L_1FQ) & \to & (P_1,FP_1) & \to & (P_o,FP_o) & \to & (Q,FQ) & \to 0 \\ & & \| & & \| & & \| & & \| & \\ & & T(0,L_1FQ) & & T(P_1,0) & & T(P_o,0) & & T(Q,0) & \end{array}$$

shows that $pd_{\underline{\underline{M}}}T(\mathbb{Q},0) = 2$; thus gl. dim $\underline{\underline{M}} = 2$, by Theorem 4.20(3).

Examples 4.27. We shall now give an example of a finite dimensional algebra $\Lambda = \begin{pmatrix} R & 0 \\ M & S \end{pmatrix}$ such that left $FPD(\Lambda) = 0 < 1 =$ left $FPD(R)$. Thus, in general, Theorem 4.20(2) cannot be strengthened.

Let $R = T_2(k)$, that is, the ring of 2×2 lower triangular matrices over a field k. We note that R has a simple right module M ($M \cong k$ as a k-module) such that, if $Tor_i^R(M,X) = 0$ for all $i > 0$, then X is a left projective R-module. Let S be the ring of matrices of the form

$$\begin{pmatrix} a & c & d \\ 0 & b & 0 \\ 0 & 0 & a \end{pmatrix}$$

with entries in k (See [Nakayama; 52]). Using duality with respect to the ground field k, one can verify that S has a simple left injective module M' ($M' \cong k$ as a k-module) such that $pd_S M' = \infty$ and such that (necessarily) $Hom_S(M',S) = 0$. Finally, we assert that there is an S-R bimodule M satisfying

a) M is simple both as a left S-module and as a right R module.

b) $pd_S M = \infty$ and $id_S M = 0$.

c) If $Tor_i^R(M,X) = 0$ for $i > 0$, then X is projective as a left R-module.

Moreover, it is easy to verify

d) left FPD(R) = gl. dim R = 1 and left FPD(S) = 0 (each left principal indecomposable S-module is uniserial of length 2).

Using once again the identification

$$_{\Lambda}\underline{\underline{Mod}} \cong \underline{Map}(F_{R}\underline{\underline{Mod}}, {}_{S}\underline{\underline{Mod}}), \quad \text{where } F = M\otimes_R -,$$

suppose $\alpha = (0,F) : (0,FA) \longrightarrow (A,B)$ represents a left Λ-module with $pd(\alpha) \leq 1$. Then there is a Λ-exact sequence

$$\begin{array}{ccccccccc}
 & & (0,FP_1) & \longrightarrow & (0,FP_o) & \longrightarrow & (0,FA) & \longrightarrow & 0 \\
 & & \downarrow & & \downarrow & & \downarrow \alpha & & \\
0 & \longrightarrow & (P_1,FP_1\oplus Q_1) & \longrightarrow & (P_o,FP_o\oplus Q_o) & \longrightarrow & (A,B) & \longrightarrow & 0 \\
 & & \| & & \| & & & & \\
 & & T(P_1,Q_1) & & T(P_o,Q_o) & & & &
\end{array}$$

where P_i is left R-projective and Q_i is left S-projective for $i = 0,1$. A diagram chase shows that the maps $\alpha = (0,f)$ and $FP_1 \to FP_o$ are monic. In particular, $Tor_i^R(M,A) \cong L_iF(A) = 0$ for all $i \geq 1$ (Recall gl. dim R = 1). It follows that A is R-projective and that $FA = M \otimes_R A$ is S-injective. The preceding facts show that α is isomorphic with $T(A,0) \oplus T(0,\text{cok } f)$. Since left FPD(S) = 0 and $pd_{\Lambda}(0, \text{cok } f) = pd_S \text{ cok } f$ (hence, cok f is left S-projective), then α is necessarily projective as a left Λ-module. It is now clear that left $FPD(\Lambda) = 0 < 1 =$ left FPD(R).

4.C. The finitistic projective dimension of $R \ltimes M$ when R is a commutative ring and M is a symmetric bimodule.

In Part B of this section, we concerned ourselves with finite projective dimension of trivial extensions of the form $\Lambda = \begin{pmatrix} R & 0 \\ M & S \end{pmatrix}$. In this case, the $R \times S$ bimodule M is "highly" non symmetric (provided R,M and S are nontrivial). In this part we shall give a brief account of the other "extreme". Thus, we shall assume throughout part C that R is a commutative ring and that M is a symmetric bimodule.

Theorem 4.28. Let R be a commutative ring and let $M \neq 0$ be a symmetric R-module. Consider the conditions:

i) *If* $\mathrm{Tor}_i^R(M,X) = 0$ *for all* $i > 0$, *then* $\mathrm{Hom}_R(N,X) = 0$ *for all submodules* N *of* M.

ii) $M \cong R/(\mathrm{ann}_R M) \oplus N$.

If M *satisfies either condition* i) *or condition* ii), *then* $\mathrm{FPD}(R \ltimes M) = \sup\{\mathrm{pd}_R Y < \infty : \mathrm{Tor}_i^R(M,Y) = 0$ *for all* $i > 0\}$ *and* $\mathrm{gl.\,dim}(R \ltimes M) = \infty$.

In particular, if S *is an* $R \ltimes M$-*module having finite projective dimension, then*

$$\mathrm{pd}_{R \ltimes M} S = \mathrm{pd}_R(R \otimes_{R \ltimes M} S) \quad \text{and} \quad \mathrm{Tor}_i^R(M, R \otimes_{R \ltimes M} S) = 0$$

for $i > 0$.

Proof. As usual, we shall identify the categories $\underline{\underline{\mathrm{Mod}}}(R \ltimes M)$ and $(\underline{\underline{\mathrm{Mod}}}\ R) \ltimes F$, where $F = M \otimes_R -$. In addition, $C : (\underline{\underline{\mathrm{Mod}}}\ R) \ltimes F \to \underline{\underline{\mathrm{Mod}}}\ R$ denotes the cokernel functor (See Section 1 and Part A of this section). Observation 4.5(2) will play an important role, especially the exact sequence

$$0 \longrightarrow H_n^F(\alpha) \longrightarrow F^{n+1}(\mathrm{cok}\ \alpha) \xrightarrow{\lambda_n} F^n A,$$

where $\alpha : FA \longrightarrow A$ represents an object in $\underline{\underline{\mathrm{Mod}}}\ R \ltimes F$ (See definition preceding Observation 4.5).

We shall first show: If $\alpha : FA \longrightarrow A$ represents an $R \ltimes M$-module and if $\mathrm{pd}_{R \ltimes M}(\alpha) = m < \infty$, then $L_n C(\alpha) = 0$ for all $n \geq 1$. Of course this statement is clear for $n = 0$. Hence, we proceed by induction and assume $n \geq 1$. There is an $R \ltimes M$ exact sequence

$$\begin{array}{ccccccccc}
 & & FB & \longrightarrow & FP \oplus F^2P & \longrightarrow & FA & \longrightarrow & 0 \\
 & & \downarrow \beta & & \searrow = & & \downarrow \alpha & & \\
0 & \longrightarrow & B & \longrightarrow & P \oplus FP & \longrightarrow & A & \longrightarrow & 0
\end{array}$$

with P an R-free module and $\epsilon : P \longrightarrow A$ an epimorphism. Our induction hypothesis gives $L_n C(\beta) = 0$ for $n \geq 1$. Hence, it suffices to show $L_1 C(\alpha) \cong H_1^F(\alpha) = 0$ (See Theorem 4.6(1)). The Snake Lemma applied to the above diagram yields an R-exact sequence

$$F^2P \xrightarrow{F\alpha \cdot F^2\epsilon} \ker \alpha \longrightarrow \mathrm{cok}\ \beta \longrightarrow P \longrightarrow \mathrm{cok}\ \alpha \longrightarrow 0.$$

Since F is right exact and ϵ is an epimorphism, the image $(F\alpha\, F^2\epsilon) =$ image $(F\alpha)$. Hence, we obtain the R-exact sequence

$$(*) \quad 0 \longrightarrow H_1^F(\alpha) \longrightarrow \operatorname{cok} \beta \longrightarrow P \longrightarrow \operatorname{cok} \alpha \longrightarrow 0$$

Theorem 4.6(2) and the fact that $L_2C(\alpha) \cong L_1C(\beta) = 0$ show that

$$\ker (F^2(\operatorname{cok} \alpha) \xrightarrow{\lambda_2} FA) \cong H_2^F(\alpha) = 0,$$

that is λ_1 is monic. Hence, we now have an R-exact sequence (see Observation 4.5(2))

$$(**) \quad 0 \longrightarrow H_1^F(\alpha) \xrightarrow{h} F(\operatorname{cok} \alpha) \xrightarrow{\lambda_1} A,$$

where $F(h) = 0$ (note $F(\operatorname{cok} \alpha) = M \otimes_R (\operatorname{cok} \alpha)$).

Now suppose condition (i) holds for M. The epimorphism $P \to \operatorname{cok}$ above gives an epimorphism $\coprod M$ (direct sum) $\cong M \otimes P \longrightarrow M \otimes (\operatorname{cok} \alpha) \supseteq H_1^F(\alpha)$. Hence, there is a submodule $N \subseteq \coprod M$ and an epimorphism $N \longrightarrow H_1^F(\alpha) \subseteq \operatorname{cok} \beta$. From Theorem 4.9(a), $\operatorname{Tor}_i^R(M, \operatorname{cok} \beta) = 0$ for all $i \geq 1$. Thus, the hypothesis of condition (i) gives

$$H_1^F(\alpha) \cong L_1C(\alpha) = 0.$$

Assume condition (ii) holds. Observe that $H_1^F(\alpha)$ and $F(\operatorname{cok} \alpha) = M \otimes_R (\operatorname{cok} \alpha)$ are naturally modules over $R/(\operatorname{ann}_R M)$. Since

$$M \cong R/(\operatorname{ann}_R M) \oplus N$$

it follows that $F(h) = 0$ in (**) if and only if $h = 0$. Thus $L_1C(\alpha) \cong H_1^F(\alpha) = 0$ in this case also.

If $\operatorname{pd}_{R \ltimes M}(\alpha) = n < \infty$, we may now apply Theorem 4.9 to obtain

$$\operatorname{pd}_{R \ltimes M}(\alpha) = \operatorname{pd}_R(\operatorname{cok} \alpha) \quad \text{and}$$

$$L_iF(\operatorname{cok} \alpha) = \operatorname{Tor}_i^R(M, \operatorname{cok} \alpha) = 0, \text{ for } i > 0.$$

Furthermore, it is elementary to show that the cokernel functor C corresponds to the functor $R \otimes_{R \ltimes M} -$ under the category isomorphism ($\underline{\text{Mod}}$ R) $\ltimes$ F $\cong$ $\underline{\text{Mod}}$ (R $\ltimes$ M). QED.

Corollary 4.29. *Let R be a commutative Noetherian ring and let $N \neq 0$ be a finitely generated R-module.*

a) *If M satisfies either of the conditions in Theorem 4.28, then* $\sup\{\operatorname{pd}_R X < \infty : \operatorname{Tor}_i^R(M,X) = 0 \text{ for } i > 0\} = \operatorname{FPD}(R)$.

b) *If N is a finitely generated R-module of finite*

projective dimension and if $\mathrm{Tor}_i^R(M,N) = 0$ for $i > 0$, then $\mathrm{pd}_R N \leq \mathrm{depth}_R M$ (Here M is an arbirary finitely generated R-module.

Proof. We refer the reader to Section 5 (Lemma 5.1) to observe that $\dim(R \ltimes M) = \dim(R)$ (Krull dimension). Now Raynaud and Gruson [64] have shown $\dim R = \mathrm{FPD}(R)$ (Hence the same statement holds for $R \ltimes M$). Thus, Theorem 4.28 yields the desired equality in part a).

From Lemma 5.1 (Section 5)

$$\mathrm{depth}\,(R \ltimes M) = \min\,(\mathrm{depth}\,R,\ \mathrm{depth}\,{}_R M)$$

By Lemma 4.4 (part A), we have $\mathrm{pd}_R N = \mathrm{pd}_R N = \mathrm{pd}_{R \ltimes M}(TN)$, where T is the tensor functor, $T:\underline{\underline{\mathrm{Mod}}}\ R \longrightarrow \underline{\underline{\mathrm{Mod}}}(R \ltimes M)$. Moreover TN is also finitely generated. Hence $\mathrm{pd}_{R \ltimes M}(TN) \leq \mathrm{depth}(R \ltimes M) \leq \mathrm{depth}_R M$. QED

Remark. If R is a commutative ring, then any nonzero cyclic R-module satisfies condition ii) of Theorem 4.28. If x is a regular nonunit of R, then R/xR satisfies condition i) of Theorem 4.28. If R is a Noetherian integral domain with field of quotionts Q, then the modules Q/R and $\coprod_{p \in \mathrm{Spec}\,R} R/p$ also satisfy i) of 4.28 (See Examples 4.30 and 4.31). We further remark that the class of R-modules satisfying condition i) of 4.28 is closed under arbitrary direct sums. Finally, we do not know of a commutative ring R and a faithful R-module M for which the conclusions of Theorem 4.28 do not hold.

Example 4.30. Let R be a regular local ring of dimension $n > 1$ with field of quotients Q such that $\mathrm{pd}_R Q < n$ (e.g., if Q is a countably generated R-module, then $\mathrm{pd}_R Q = 1$. See Osofsky; [54].). Then $R \ltimes (Q/R)$ is a local (non Noetherian) ring such that

$$\mathrm{FPD}(R \ltimes Q/R) = \sup\{\mathrm{pd}_R X : X \text{ is } R\text{-torsion free}\} <$$

$$n = \dim(R \ltimes Q/R) = \mathrm{FPD}(R).$$

Proof. If X is an R-module such that $\mathrm{Tor}_i^R(Q/R,X) = 0$ for $i \geq 1$, then X is necessarily torsion free and so $\mathrm{Hom}_R(N,X) = 0$, for all $N \subseteq Q/R$ (torsion). It remains only to show that $\mathrm{pd}_R X < n$, whenever X is torsion free. In this case, there is an monomorphism $X \xrightarrow{f} \coprod Q$ (injective envelope of X). But $\mathrm{pd}_R \coprod Q = \mathrm{pd}_R Q < n$. Hence if $\mathrm{pd}_R X = n$, then $pd_R(\text{cok } f) = n + 1 > \text{gl. dim } R$. Thus

$$\sup\{pd_R X : X \text{ is } R\text{-torsion free}\} < n.$$

Let $\mathbf{F} = Q/R \otimes_R -$ and recall the isomorphism of categories $\underline{\text{Mod}}(R \ltimes M) \cong (\underline{\text{Mod}}\ R) \ltimes \mathbf{F}$. It is easy to observe that $L_i F^j = 0$ for $i + j \geq 3$, since $L_i F = 0$ for $i \geq 2$ and $F^2 = 0$. However, gl. dim $(R \ltimes Q/R) = \infty$ even though gl. dim $R < \infty$. Thus the assumption in Theorem 4.14 that $F(P)$ be F^r-acyclic for all projectives P and $r \geq 1$ cannot simply be dropped. In the above situation,

$$F(R) = Q/R \text{ and } \text{Tor}_1^R(Q/R, Q/R) \cong Q/R \neq 0.$$

The verification of assertions made in the following example is similar to that in 4.30.

Example 4.31. Let R be a commutative integral domain (not a field) and let $M = \coprod_{p \in \text{Spec } R} (R/p)$. Then

$$\text{FPD}(R \times M) = \sup\{pd_R X < \infty : X \text{ is } R\text{-flat}\}.$$

4.D. <u>The injective dimension of</u> $R \ltimes M$ <u>as a</u> (left) <u>module</u>.

In Section 5 of our paper, we shall exploit the situation in which $id_{R \ltimes M}(R \ltimes M) < \infty$ to develope a theory of Gorenstein modules over commutative rings from the point of view of Bass' paper [12]. In this brief space we lay some of the ground work for Section 5 and, in addition, we make some general remarks as to when $id_{R \ltimes M}(R \ltimes M) < \infty$.

Throughout our discussion we shall make use of the category isomorphism $_{R \ltimes M}\underline{\underline{\text{Mod}}} \cong {_R\underline{\underline{\text{Mod}}}} \ltimes F \cong G \rtimes {_R\underline{\underline{\text{Mod}}}}$, where $F = M \otimes_R -$ and $G = \text{Hom}_R(M,-)$ (See Section 1). The preceding notation will be standard in what follows.

<u>Theorem</u> 4.32. <u>Let</u> R <u>be a ring and let</u> M <u>be an</u> R-<u>bimodule</u> <u>such that</u> (as a left R-module)

$$\text{Ext}_R^i(M,M) \cong \begin{cases} R & \text{if } i = 0 \\ 0 & \text{if } i > 0. \end{cases}$$

<u>Then</u> (left) $id_{R \ltimes M}(R \ltimes M) = $ (left) $id_R M$.

Proof. From Lemma 4.4 (Part A), the object

$$H(M) = \begin{array}{c} GM \oplus M \\ \searrow = \\ G^2M \oplus GM \end{array}$$

has $R \ltimes M$-injective dimension equal to $id_R M$. Under the category isomorphism ${}_R\underline{\underline{Mod}} \ltimes F \cong G \rtimes {}_R\underline{\underline{Mod}}$, the object

$$R \ltimes M = T(R) = \begin{array}{c} FR \oplus F^2R \\ \searrow = \\ R \oplus FR \end{array}$$

corresponds to the object

$$\begin{array}{c} R \oplus FR \\ \searrow \tau \\ GR \oplus GFR \end{array}$$

where $\tau : R \longrightarrow GFR = Hom_R(M,M)$ is the natural ring homomorphism. However, by assumption τ is an isomorphism,

$$GR \cong Hom_R(M,R) \cong Hom_R(M, Hom_R(M,M)) \cong G^2R$$

and $FR = M$. Thus, the above object is isomorphic to $H(M) \in G \rtimes {}_R\underline{\underline{Mod}}$ and hence $id_{R \ltimes M}(R \ltimes M) = id_R M$.

Corollary 4.33. [Gulliksen (Artin case); 30]. a) *If R is a (commutative) complete local ring with maximal ideal m and if E is the injective envelope of R/m, then $R \ltimes E$ is a self injective ring.*

b) *If Λ is a finite algebra* (finitely generated as a module) *over the commutative Artin ring k, then $\Lambda \ltimes \Lambda^d$ is quasi-Frobenius, where $\Lambda^d = Hom_k(\Lambda, E)$* ($E$ is the injective k-module giving a perfect contravariant duality between k-projectives and k-injectives).

Proof. a) Of course $Ext_R^i(E,E) = 0$ for $i > 0$ and it is well-known that $Hom_R(E,E) \cong R$ (See Matlis [42]).

b) We have that Λ^d is a finitely generated injective (left or right) Λ-module and that $Hom\,(\Lambda^d, \Lambda^d) \cong \Lambda$ (as a left or right module).

QED

In general, it is not true that $id_{R \ltimes M}(R \ltimes M) = id_R M$ (even when both are finite). For example, let $\Lambda = \begin{pmatrix} k & 0 \\ k & k \end{pmatrix}$, where k is a field. Then $\Lambda = (k \times k) \ltimes k$ (k a nonsymmetric bimodule over $k \times k$) and

$\mathrm{id}_\Lambda \Lambda = 1 > 0 = \mathrm{id}_{k \times k}(k)$. An explanation of the general lack of equality of the dimensions $\mathrm{id}_{R \ltimes M}(R \ltimes M)$ and $\mathrm{id}_R M$ goes as follows. If M satisfies the hypothesis of Theorem 4.32, then an (left) R-injective resolution of M can be "lifted" via the functor

$$H : {}_R\underline{\underline{\mathrm{Mod}}} \longrightarrow G \rtimes {}_R\underline{\underline{\mathrm{Mod}}}$$

to an injective resolution of $H(M)$ in $G \rtimes {}_R\underline{\underline{\mathrm{Mod}}}$. It follows that $R^iK(H(M)) = 0$ for $i > 0$, where $K: G \rtimes {}_R\underline{\underline{\mathrm{Mod}}} \longrightarrow {}_R\underline{\underline{\mathrm{Mod}}}$ is the kernel functor (See Lemma 4.4). In more concrete terms, the preceding statement says that $\mathrm{Ext}^i_{R \times M}(R, R \ltimes M) = 0$ for $i > 0$ (We are using the fact that $R \ltimes M$ corresponds to $H(M)$ under ${}_R\underline{\underline{\mathrm{Mod}}} \ltimes F \cong G \rtimes {}_R\underline{\underline{\mathrm{Mod}}}$). However, in the above example, $\mathrm{Ext}^1_\Lambda(k \times k, \Lambda) \neq 0$ where $k \times k$ is considered as a Λ-module. The condition $\mathrm{Ext}^i_{R \ltimes M}(R, R \ltimes M) = 0$ for $i > 0$ will receive more attention in Proposition 4.35.

Proposition 4.34. *Let* R *be a ring and let* M *be an* R-*bimodule. Then* (left) $\mathrm{id}_R M \leq$ (left) $\mathrm{id}_{R \ltimes M}(R \ltimes M)$.

Proof. As noted in the proof of Theorem 4.32, the module $R \times M$ corresponds to the object

$$\begin{array}{ccc} R \oplus FR & & R \oplus M \\ \searrow\tau & = & \searrow\tau \\ GR \oplus GFR & & \mathrm{Hom}(M,R) \oplus \mathrm{Hom}(M,M). \end{array}$$

The conclusion now follows from Lemma 4.2. QED.

Proposition 4.35. *Let* R *be a ring and let* M *be an* R-*bimodule. If as left modules* $\mathrm{Ext}^i_{R \ltimes M}(R, R \ltimes M) = 0$ *for all* $i > 0$, *then the following statements hold for the left* R-*modules* M *and* B = right ann_R

1) $\mathrm{id}_R(B \oplus M) = \mathrm{id}_{R \ltimes M}(R \ltimes M)$.

2) *The natural map* $R \xrightarrow{\tau} \mathrm{Hom}_R(M,M)$ *is an epimorphism.*

3) $\mathrm{Ext}^i_R(M, M \oplus B) = 0$ *for all* $i > 0$.

4) $\mathrm{Hom}_R(M,B) = 0$.

5) *If also* $\mathrm{Ext}^i_{R \ltimes M}(\mathrm{Hom}_R(M,M), R \ltimes M) = 0$ *for all* $i > 0$,

then B is a left direct summand of R.

Proof. We shall present this proof within the framework of the category $G \rtimes {}_R\underline{\underline{\text{Mod}}}$ (as defined in Section 1), where $G = \text{Hom}_R(M,-) : {}_R\underline{\underline{\text{Mod}}} \longrightarrow {}_R\underline{\underline{\text{Mod}}}$. The kernel functor $K : G \quad {}_R\underline{\underline{\text{Mod}}} \longrightarrow {}_R\underline{\underline{\text{Mod}}}$ (See Section 1) is easily seen to be naturally equivalent to the functor $\text{Hom}_{R \ltimes M}(R,-)$. In the category $G \rtimes {}_R\underline{\underline{\text{Mod}}}$, the module $R \ltimes M$ is represented by the map

$$\rho = \begin{array}{c} R \oplus M \\ \searrow^{\tau} \\ \text{Hom}_R(M,R) \oplus \text{Hom}_R(M,M) \end{array} = \begin{array}{c} R \oplus M \\ \searrow \\ GR \oplus GM \end{array},$$

where $\tau : R \longrightarrow \text{Hom}_R(M,M)$ is the natural map. Note that $K(\rho) = B \oplus M$. Hence, parts 1) and 3) are immediate consequences of Theorem 4.11, since necessarily $R^iK(\rho) = 0$ for all $i > 0$. Furthermore, the dual statement of Theorem 4.6(1) with respect to $G \rtimes {}_R\underline{\underline{\text{Mod}}}$ and K gives

$$0 = R^1K(\rho) \cong \ker G(\rho)/\text{image } \rho.$$

But this statement implies that τ is an epimorphism and that $G(\tau)$ is a monomorphism. Now $G(\tau)$ is the natural map

$$\text{Hom}_R(M,R) \longrightarrow \text{Hom}_R(M,\text{Hom}_R M,M))$$

with kernel $\text{Hom}_R(M,R) \longrightarrow \text{Hom}_R(M,R)$. Therefore, $\text{Hom}_R(M,B) = 0$ and both 2) and 4) have been established. At this point, we also have an exact sequence of left R-modules $0 \to B \to R \xrightarrow{\tau} \text{Hom}_R(M,M) \to 0$. Since

$$R^i\text{Hom}_{R \ltimes M}(R, R \ltimes M) \cong \text{Ext}^i_{R \ltimes M}(R, R \ltimes M) = 0 \quad \text{for} \quad i > 0,$$

the natural isomorphism

$$\text{Hom}_{R \ltimes M}(N, R \ltimes M) \cong \text{Hom}_R(N, \text{Hom}_{R \ltimes M}(R, R \ltimes M)),$$

for all left R-modules N, gives natural isomorphisms of right derived functors

$$\text{Ext}^i_{R \ltimes M}(N, R \ltimes M) \cong \text{Ext}^i_R(N, \text{Hom}_{R \ltimes M}(R, R \ltimes M)).$$

Moreover,

$$\text{Hom}_{R \ltimes M}(R, R \ltimes M) \cong K(\rho) = B \oplus M.$$

Hence, the hypothesis of 4) implies $\text{Ext}^1_R(\text{Hom}_R(M,M),B) = 0$ and, therefore, the exact sequence $0 \to B \to R \xrightarrow{\tau} \mathrm{Hom}_R(M,M) \to 0$ splits.

QED.

Corollary 4.36. *Let* R *be a ring and let* M *be an* R-*bimodule Then* $R \ltimes M$ *is left self injective if and only if*

1) M *and* $B = \text{rt. ann}_R M$ *are injective left* R-*modules*,

2) *The natural map* $R \xrightarrow{\tau} \mathrm{Hom}_R(M,M)$ *is an epimorphism, and*

3) $\mathrm{Hom}_R(M,B) = 0$.

Proof. The necessity follows directly from Proposition 4.35.

The sufficiency follows from the observation that the object

$$\begin{array}{c} R \oplus M \\ \searrow \tau \\ GR \oplus GM \end{array}$$

in $G \rtimes {}_R\mathrm{Mod}$ $(G = \mathrm{Hom}_R(M,-))$, corresponding to the left module $R \times M$, is isomorphic with the direct sum of the objects

$$\begin{array}{c} GM \oplus M \\ \searrow = \\ G^2M \oplus GM \end{array} \qquad \text{and} \qquad \begin{array}{c} GB \oplus B \\ \searrow = \\ G^2B \oplus GB \end{array} = \begin{array}{c} B \\ \downarrow \\ 0 \end{array}$$

(note $GB \cong G^2B = 0$). From Corollary 1.6 it follows that $R \times M$ is left self injective.

QED.

Corollary 4.37. *Let* R *be a local* (not necessarily commutative) *ring and let* $M \neq 0$ *be an* R-*bimodule. Then* $R \ltimes M$ *is left self injective if and only if* M *is left* R-*injective and* $R \cong \mathrm{Hom}_R(M,M)$.

QED.

Section 5. Gorenstein modules

Recently Sharp [66,67,68,69], Foxby [23,24] and Herzog and Kunz [34] have developed a theory of Gorenstein modules (called canonical modules in [34]) over a commutative Noetherian ring which generalizes the work of Bass [11,12] on commutative rings which are locally of finite self injective dimension. In [63] Reiten (and also independently Foxby [23]) established a connection between Gorenstein rings and the existence of Gorenstein modules as follows: If A is a Cohen-Macaulay ring and $M \neq 0$ a finitely generated (symmetric) A-module, then $A \times M$ is a Gorenstein ring if and only if M is a Gorenstein module of rank one. It is this basic result which provides the essential motivation to what follows. In this section, we shall extend some of the results of the aforementioned authors on Gorenstein modules of rank one and, in addition, offer some direction in the study of Gorenstein modules of larger rank over local rings. Using an example of Ferrand and Raynaud [18], we establish the existence of a Cohen-Macaulay local ring which does not have a Gorenstein module of any rank. Perhaps more importantly however, we demonstrate how the theory of trivial extensions combined with the original work of Bass [12] yields most of the basic results concerning Gorenstein modules of rank one.

Throughout this section, we shall consider only trivial extensions of commutative Noetherian rings by finitely generated symmetric modules (hence, the trivial extensions will always be commutative and Noetherian). Our terminology with respect to commutative algebra will be standard. Perhaps we should mention, however, that we shall use the term "depth" rather than "codimension" as in Bass [12] or "Grade" as in Kaplansky [36]. We take this opportunity to thank R.Sharp and H.-B. Foxby for many informative conservations on the subject matter which follows.

In our first lemma, we compile several elementary facts concerning trivial extensions of commutative rings by finitely generated (symmetric) modules. The proof is straightforward and we omit it. We mention that part (iv) of this lemma is taken from Foxby [23] and that Sharp pointed out part (v) to the authors.

Lemma 5.1. *Let* A *be a commutative Noetherian ring and let* $M \neq 0$ *be a finitely generated* A-*module*. *The following statements hold for the ring* $A \times M$.

i) *The prime ideals in* $A \ltimes M$ *are of the form*

$\underline{p} \times M = \{(x,m) \in A \ltimes M\colon x \in \underline{p},\ m \in M\}$ *where* $\underline{p}$ *is a prime ideal of* A.

ii) *If* $\underline{p}$ *is a prime ideal of* A, *then* $ht_{A \ltimes M}(\underline{p} \times M) = ht_A(\underline{p})$.

iii) $\dim(A \ltimes M) = \dim A$.

iv) *If* S *is a multiplicatively closed set in* A, *then* $S \times M$ *is a multiplicatively closed set in* $A \ltimes M$ *and*

$$(r,x)/(s,y) \longmapsto (r/s,(sx-ry)/s^2)$$

defines an isomorphism

$$(S \times M)^{-1}(A \ltimes M) \longrightarrow S^{-1}A \ltimes S^{-1}M.$$

In particular, $(A \ltimes M)_{\underline{p} \times M} = A_{\underline{p}} \ltimes M_{\underline{p}}$, *for* $\underline{p}$ *a prime ideal of* A.

v) $\operatorname{depth}(A \ltimes M) = \operatorname{depth}_A(A \oplus M) = \min(\operatorname{depth} A, \operatorname{depth}_A M)$.

vi) *If* N *is a finitely generated* A-*module, then* $\operatorname{depth}_{A \ltimes M}(N) = \operatorname{depth}_A N$.

Our next two theorems form the cornerstone on which much of our development of the theory of Gorenstein modules depends. The proofs of Theorem 5.2 and Theorem 5.4 require an application of our results in Section 4(D). We remark, however, that the result in Section 4(D), necessary for this section, could be obtained independently from most of the previous machinery, if one wished to do so.

Theorem 5.2. *Let* A *be a commutative Noetherian ring and let* $M \neq 0$ *be a finitely generated* A-*module. If* X *is a finitely generated* $A \ltimes M$-*module of finite* $A \ltimes M$-*injective dimension* n, *then the following statements are true*.

1) $\overline{X} = \operatorname{Hom}_{A \ltimes M}(A,X)$ *is a nonzero finitely generated* A-*module with* $\operatorname{id}_A \overline{X} = n$.

2) *If* $X \longrightarrow I^{\bullet}$ *is a (minimal)* $A \ltimes M$-*injective resolution*

<u>of</u> X, <u>then</u> $\overline{X} \longrightarrow \mathrm{Hom}_{A \ltimes M}(A, I^{\bullet})$ <u>is a</u> (minimal) A-<u>injective resolution of</u> $\overline{X}$.

3) <u>If</u> N <u>is an</u> A-<u>module, then there is a natural isomorphism</u> $\mathrm{Ext}^i_{A \ltimes M}(N,X) \cong \mathrm{Ext}^i_A(N,\overline{X})$, <u>for all</u> $i \geq 0$.

4) <u>If</u> A <u>is a local ring, then</u> $n = \mathrm{depth}\,(A \ltimes M) = \mathrm{depth}\,A = \mathrm{depth}_A M$.

Proof. In view of the localization procedure, as described in Lemma 5.1(iv), it suffices to prove the above statements when A is a local ring (which we henceforth assume to be the case). From Bass [Lemma 3.3; 12] and Lemma 5.1(v), (vi), we have

$$n = \mathrm{id}_{A \times M} X = \mathrm{depth}(A \times M) \leq \mathrm{depth}_{A \times M}(A) = \mathrm{depth}\ A.$$

By Kaplansky [Theorem 2.19; 36], depth $A = n$ and, for the same reasons, $\mathrm{depth}_A M = n$. Hence, by Kaplansky [Theorem 2.17; 36], it follows that $\mathrm{Ext}^i_{A \times M}(A,X) = 0$, for $i > 0$. The general construction of the ring $A \times M$ easily gives that $\overline{X} = \mathrm{Hom}_{A \times M}(A,X)$ is a nonzero finitely generated A-module. Furthermore, a standard "change of rings" argument gives parts 2) and 3) (in view of $\mathrm{Ext}^i_{A \times M}(A,X) = 0$ for $i > 0$) and that $\mathrm{id}_A \overline{X} \leq n$. By Bass [Lemma 3.3; 12], $\mathrm{id}_A \overline{X} = \mathrm{depth}\ A = n$. Finally, part 4) has already been established in the preceding discussion. QED.

The next lemma is a consequence of a standard "change of rings" argument (as can be found in Kaplansky [36]) and we omit its proof.

<u>Lemma</u> 5.3. <u>Let</u> A <u>be a local ring and let</u> $M \neq 0$ <u>be a finitely generated</u> A-<u>module such that</u> $\mathrm{Ext}^i_A(M,M) = 0$, <u>for all</u> $i > 0$. <u>If</u> $\mathfrak{X}$ <u>is both an</u> A-<u>sequence and an</u> M-<u>sequence, then the following statements hold for</u> M <u>and</u> $\mathfrak{X}$.

i) $\mathfrak{X}$ <u>is a</u> $\mathrm{Hom}_A(M,M)$-<u>sequence. Hence</u> $\mathrm{depth}_A \mathrm{Hom}_A(M,M) \geq \min\,(\mathrm{depth}\ A,\ \mathrm{depth}_A M)$.

ii) $\mathrm{Ext}^i_{A/\mathfrak{X}A}(M/\mathfrak{X}M,\ M/\mathfrak{X}M) = 0$, <u>for</u> $i > 0$.

iii) $\mathrm{End}_{A/\mathfrak{X}A}(M/\mathfrak{X}M) \cong \mathrm{End}_A M/\mathfrak{X}\,\mathrm{End}_A M$.

QED.

Among other features, the following theorem reduces the study of Gorenstein rings of the form $A \times M$ to those where A has connected prime spectrum and $\mathrm{ann}_A M = 0$.

Theorem 5.4. Suppose A is a commutative Noetherian ring and $M \neq 0$ is a finitely generated A-module such that $A \times M$ is a Gorenstein ring. Then

(a) A is Cohen-Macaulay,

(b) $\mathrm{Ext}_A^i(M,M) = 0$ for $i > 0$,

(c) $A \cong \mathrm{ann}_A M \times \mathrm{End}_A M$ (ring product),

(d) $B = \mathrm{ann}_A M$ is a Gorenstein ring.

Proof. We first note there is an exact sequence of A-modules,

$$0 \longrightarrow \mathrm{ann}_A M \longrightarrow A \xrightarrow{\tau} \mathrm{End}_A M.$$

We shall prove, for $\underline{p} \in \mathrm{Spec}\ A$, that $\tau_{\underline{p}}$ is an isomorphism if $\underline{p} \in \mathrm{Supp}\ M$ and, of course, zero if $p \notin \mathrm{Supp}\ M$. Assuming we keep this promise, we observe that part (c) necessarily follows. In addition, for each $\underline{p} \in \mathrm{Spec}\ B = \mathrm{Spec}\ A - \mathrm{Supp}_A M$, it follows from Lemma 5.1 (iv) that $B_{\underline{p}} \cong (A \ltimes M)_{p \times M} \cong A_{\underline{p}}$ is a Gorenstein ring. Finally, it suffices to establish parts (a) and (b) locally at a prime $\underline{p} \in \mathrm{Spec}\ A$.

In view of the preceding discussion, we shall henceforth assume that A is a local ring and that $M \neq 0$ is a finitely generated A-module such that $A \ltimes M$ is a Gorenstein ring. By [Bass (Theorem 4.1); 12], $A \ltimes M$ is necessarily a Cohen-Macaulay ring and $\mathrm{id}_{A \ltimes M}(A \ltimes M) = n = \dim(A \ltimes M)$. We now combine the conclusions of Lemma 5.1 and Theorem 5.2 with the fact

$$\mathrm{Hom}_{A \ltimes M}(A,\ A \ltimes M) = M \oplus B,\ B = \mathrm{ann}_A M,$$

to obtain the equalities

$$\begin{aligned} n &= \dim A = \dim A \ltimes M = \mathrm{depth}\ A \ltimes M \\ &= \mathrm{depth}\ A = \mathrm{depth}_A M = \mathrm{id}_A M. \end{aligned}$$

In particular, A is necessarily a Cohen-Macaulay ring. As in the proof of Theorem 5.2, we apply [Kaplansky (Theorem 217); 36] to obtain $\mathrm{Ext}^i_{A \ltimes M}(A, A \ltimes M) = 0$ for $i > 0$. For the same reason,

$$\mathrm{Ext}^i_{A \ltimes M}(M, A \ltimes M) = \mathrm{Ext}^i_A(M, M \oplus B) = 0$$

for $i > 0$, (Theorem 5.2(3)) and $\mathrm{Ext}^i_{A \ltimes M}(\mathrm{Hom}_A(M,M), A \ltimes M) = 0$, for $i > 0$, (Lemma 5.3 (i)). We now call upon Proposition 4.35 (Section 4(D)) in order to establish that $\tau : A \longrightarrow \mathrm{End}_A M$ is an epimorphism and that $B = \mathrm{ann}_A M$ is a direct summand of A. Since A is local and $\mathrm{End}_A M \neq 0$, it follows that $B = \mathrm{ann}_A M = 0$: Hence τ is an isomorphism. This completes our proof.

QED.

Corollary 5.5. If A is a commutative Noetherian ring and $M \neq 0$ is a finitely generated A-module such that $A \ltimes M$ is a Gorenstein ring, then the Gorenstein locus of A is

$$\{\underline{p} \in \mathrm{Spec}\, A : M_{\underline{p}} \text{ is } A_{\underline{p}}\text{-projective}\}.$$

Hence, A is a Gorenstein ring in this case if and only if M is a projective A-module. Moreover, if Spec A is connected, then $\mathrm{Supp}_A M =$ Spec A and $\mathrm{ann}_A M = 0$. QED

The notion of a Gorenstein module (or canonical module) goes back to Grothendieck (See [28; pages 94,95] and [27]) and the so-called module of dualizing differentials (also see Section 5 (Remarks of Serre) of Bass [12] and Sharp [Theorem 3.1: 68]). The construction of such modules arises in the following classical fashion: Let R be a Gorenstein local ring (e.g. a regular local ring) and let A be a Cohen-Macaulay homomorphic image of R. The A-module $\Omega = \mathrm{Ext}^d_R(A,R)$, where $d = \dim R - \dim A$, is called the module of dualizing differentials for A and it possesses the following basic properties:

1) If $R \longrightarrow I^\bullet$ is a minimal injective resolution of R, then the A-injective complex $\mathrm{Hom}_R(A, I^\bullet)_{i \geq d}$ is a minimal A-injective resolution of Ω.

2) $\mathrm{id}_A \Omega = \mathrm{depth}_A \Omega$.

3) For each $\underline{p} \in \mathrm{Spec}\, A$, $\mu_i(\underline{p}, \Omega) = \delta_{i, \mathrm{ht}\, \underline{p}}$ (Kronecker delta), where

$$\mu_i(\underline{p},\Omega) = \dim_{k(\underline{p})} \mathrm{Ext}^i_{A_{\underline{p}}}(k(\underline{p}),\Omega_{\underline{p}}) = \dim_{k(\underline{p})} \mathrm{Ext}^i_A(A/\underline{p},\Omega)_{\underline{p}}$$

($k(\underline{p})$ is the **residue** field of $A_{\underline{p}}$) is the number of the copies of $E(A/\underline{p})$ (injective envelope of $A/\underline{p}$) in the $i^{\underline{th}}$ injective in a minimal injective resolution of Ω (See Bass [12]).

Following Herzog and Kunz [34], we shall say that Ω is a canonical module (Gorenstein module of rank one in the sense of Sharp [69]) over the ring A if

$$\mu_i(\underline{p},\Omega) = \delta_{i,\mathrm{ht}\ \underline{p}}.$$

More generally, if Ω satisfies:

$$\mu_i(\underline{p},\Omega) \neq 0 \text{ if and only if } \mathrm{ht}\ \underline{p} = i,$$

then Ω will be called a Gorenstein module. Hence, a canonical module is a special type of Gorenstein module.

In the following theorem and its corollaries, we use the notion of trivial extension to demonstrate the fundamental connection between the theory of Gorenstein rings, as developed by Bass [12], and the more general theory of canonical modules. The results of this theorem are due to Sharp [67,68], Foxby [23] and Reiten [63]. One should also compare it with Satz 6.1 [Herzog and Kunz, 34].

Theorem 5.6. Let A be a commutative Noetherian ring having connected prime spectrum and let $\Omega \neq 0$ be a finitely generated A-module. Then the following statements are equivalent:

1) Ω is a canonical module.

2) $\mu_i(\underline{m},\Omega) = \delta_{i,\mathrm{ht}\ \underline{m}}$, for all maximal ideals $\underline{m}$.

3) $A \ltimes \Omega$ is a Gorenstein ring.

4) $\mathrm{id}_{A_{\underline{m}}}\Omega_{\underline{m}} < \infty$, for each maximal ideal $\underline{m}$, and

$$\mathrm{Ext}^i_A(\Omega,\Omega) = \begin{cases} A, & \text{if } i = 0 \\ 0, & \text{if } i > 0 . \end{cases}$$

5) $\mu_{\mathrm{ht}\ \underline{m}}(\underline{m},\Omega) = 1$, for each maximal ideal $\underline{m}$, and $\mathrm{Ext}^i_A(\mathrm{Ext}^j_A(M,\Omega),\Omega) = 0$, for $i < j$ and for all finitely

<u>generated modules</u> M.

6) $\mu_{\mathrm{ht}\ \underline{m}}(\underline{m},\Omega) = 1$ <u>and</u> $\mathrm{id}_{A_{\underline{m}}}\Omega_{\underline{m}} = \mathrm{depth}_{A_{\underline{m}}}\Omega_{\underline{m}}$, <u>for all maximal ideals</u> $\underline{m}$.

Proof. The usual localization procedure in commutative ring theory together with Lemma 5.1 (iv) show that it is sufficient to establish the above equivalences when A is a local ring with maximal ideal $\underline{m}$.

That 3) ===> 4) follows directly from Theorem 5.4.
The implication 4) ===> 3) is a consequence of Theorem 4.32.

Proof of 3) ===> 1). Since $\mathrm{ann}_A \Omega = 0$ (Theorem 5.4(c)), it follows that $\Omega \cong \mathrm{Hom}_{A \ltimes \Omega}(A, A \ltimes \Omega)$. Hence, the combination of [Bass Theorem 4.1(6); 12] together with Lemma 5.1 (ii) and Theorem 5.2(1), (2) gives the desired result.

Proof of 1) ===> 5). From [Bass (Lemma 3.3); 12] and the hypothesis in 1), we have $\mathrm{id}_{A_{\underline{q}}}\Omega_{\underline{q}} = \mathrm{depth}\ A_{\underline{q}} \leq \mathrm{ht}\ \underline{q}$, for each $\underline{q} \in \mathrm{Spec}\ A$.

Hence, if $i < j$ and $\mathrm{ht}\ \underline{q} = i$, we have

$$\mathrm{Ext}^i_A(M,\Omega)_{\underline{q}} = \mathrm{Ext}^i_{A_{\underline{q}}}(M_{\underline{q}},\Omega_{\underline{q}}) = 0,$$

since $j > \mathrm{ht}\ \underline{q} \geq \mathrm{id}_{A_{\underline{q}}}\Omega_{\underline{q}}$. Therefore

$$\mathrm{Hom}_A(\mathrm{Ext}^j_A(M,\Omega),\ E(A/\underline{q})) = 0,$$

if $\mathrm{ht}\ \underline{q} < j$. From the description of an injective resolution for Ω given in 1) together with the preceding statement, it now follows that $\mathrm{Ext}^i(\mathrm{Ext}^j_A(M,\Omega),\Omega) = 0$ for $i < j$ and for all finitely generated modules M.

Proof of 5) ===> 6). Let k be the residue field of A and let $n = \mathrm{depth}_A\Omega$. Then, by [Bass (Proposition 2.9); 12], $\mathrm{Ext}^i_A(k,\Omega) = 0$ for $i < n$ and $\mathrm{Ext}^n_A(k,\Omega) \neq 0$. Since $\mathrm{Ext}^{n+i}_A(k,\Omega) = k^{r_i}$ (or zero) for $i = 1,2,\ldots$ and since $\mathrm{Ext}^n_A(\mathrm{Ext}^{n+i}_A(k,\Omega),\Omega) = 0$, by hypothesis $(i > 0)$, it follows that $\mathrm{Ext}^{n+i}_A(k,\Omega) = 0$, for all $i > 0$, Therefore,

$$\operatorname{depth}_A \Omega = n = \operatorname{id}_A \Omega = \operatorname{depth} A,$$

by [Kaplansky (Theorem 212); 36] and [Bass (Lemma 3.3); 12]. By taking a sequence $\mathfrak{X}$, which is both a maximal A-sequence and a maximal Ω-sequence, one obtains an $A/\mathfrak{X}A$-injective module $\Omega/\mathfrak{X}\Omega$ [via Kaplansky (Theorem 206); 36]. By [Kaplansky (Theorem 207); 36], $A/\mathfrak{X}A$ has Krull dimension zero. Thus A is Cohen-Macaulay and $n = \operatorname{ht} \underline{m}$ (Clearly $\operatorname{Ext}_A^n(k,\Omega) \cong k$ and $\mu_{\operatorname{ht} \underline{m}}(\underline{m},\Omega) = 1$).

Proof of 6) ===> 4). Now $\operatorname{Ext}_A^i(\Omega,\Omega) = 0$ for $i > 0$ is immediate from [Kaplansky (Theorem 218); 36]. Let $\mathfrak{X}$ be a common maximal A-sequence and maximal Ω-sequence. The same "change of rings" argument as in the preceding 5) ===> 6) as well as [Bass (Corollary 2.6); 12] shows that $\Omega/\mathfrak{X}\Omega = E$, where E is the $A/\mathfrak{X}A$-injective envelope of $k = A/\underline{m}$. Of course (as in 5) ===> 6)), we also have that $A/\mathfrak{X}A$ is an Artin (= descending chain condition on ideals) local ring. Hence

$$\operatorname{End}_{A/\mathfrak{X}A}(\Omega/\mathfrak{X}\Omega) \cong \operatorname{End}_{A/\mathfrak{X}A}(E) \cong A/\mathfrak{X}A.$$

It follows that $\operatorname{End}_A\Omega \cong A$ from Lemma 5.3 and [Kaplansky (Theorem 172; 36].

That 1) ===> 2) is trivial.

That 2) ===> 6) is nearly identical to the argument used to show that 5) ===> 6) above. QED.

<u>Corollary 5.7. Suppose</u> A <u>is a local commutative ring and</u> Ω <u>is canonical</u> A-<u>module</u>. <u>Then there is a natural isomorphism</u> $\operatorname{Ext}_A^n(N,\Omega) \cong \operatorname{Hom}_A(N,E)$ <u>on all</u> A-<u>modules</u> N <u>of finite length, where</u> $n = \dim A$ <u>and</u> $E = E(k)$ <u>is the injective envelope of the residue field</u>.

Proof. The above statement is a consequence of Theorem 5.2(3), Theorem 5.6(3), [Bass (Theorem 4.1(5)); 12] and the fact

$$\Omega \cong \operatorname{Hom}_{A \ltimes \Omega}(A, A \ltimes \Omega). \qquad \text{QED.}$$

In an (unpublished) communication, Foxby has shown the statement in Corollary 5.7 to be equivalent to those in Theorem 5.6 when A is a local ring.

<u>Corollary 5.8. Suppose</u> A <u>is a local commutative ring and</u> Ω

is a canonical A-module. Then every system of paramters on Ω generates an irreducible submodule of Ω.

Proof. From Theorem 5.4, we have that A is a Cohen-Macaulay ring and, from Theorem 5.6, it is easily seen that Ω is a Cohen-Macaulay A-module of the same dimension as A. Therefore, if $x_1,\ldots,x_n$ is a system of parameters on Ω (under the preceding conditions on A and Ω), then $x_1,\ldots,x_n$ is necessarily a maximal Ω-sequence. Since $\mathrm{Hom}_A(\Omega,\Omega) \cong A$ and $\mathrm{Ext}_A^i(\Omega,\Omega) = 0$ for $i > 0$ (Theorem 5.6(4)), it is easily observed that $x_1,\ldots,x_n$ is also a maximal A-sequence (hence, also a system of parameters for A). Therefore, $(x_1,0),\ldots,(x_n,0)$ is a system of parameters for $A \ltimes \Omega$. By Theorem 5.6(3) and [Bass (Theorem 4.1(3)); 12], $(x_1,0),\ldots,(x_n,0)$ generates an irreducible ideal in $A \ltimes \Omega$ of the form $I \times N$, where N is the submodule of Ω generated by $x_1,\ldots,x_n$. Since $I \times N$ is irreducible in $A \ltimes \Omega$, it is straightforward that N is irreducible in Ω. QED.

Corollary (Reiten, Foxby, Sharp) 5.9. Suppose A is a commutative ring with finite Krull dimension and with connected prime spectrum. Then A has a canonical module if and only if A is a Cohen-Macaulay homomorphic image of a Gorenstein ring of finite Krull dimension. Thus, every Cohen-Macaulay, complete local ring has a canonical module [See Cohen; 16].

Proof. If A has a canonical module, then Theorem 5.4, Theorem 5.6(3) and Lemma 5.1 (iii) give that A is a Cohen-Macaulay homomorphic image of a Gorenstein ring of finite Krull dimension.

Now suppose that B is a Gorenstein ring of finite Krull dimension and that A is a Cohen-Macaulay homomorphic image of B. Then it is well-known that $\Omega = \mathrm{Ext}_B^d(A,B)$ is a canonical A-module, where $d = \mathrm{grade}_B A$ (See Sharp [68] or Grothendieck [28]). QED.

A further examination of the proof of Theorem 5.6 reveals that, with a few minor modifications of proof, one can actually establish the following theorem whose results are due to Sharp [67] and Foxby [23].

Theorem 5.10. Suppose A is a commutative Noetherian ring with connected prime spectrum and suppose $\Omega \neq 0$ is a finitely generated A-module. Then the following statements are equivalent.

a) Ω is a Gorenstein A-module.

b) [Sharp]. For each maximal ideal $\underline{m}$,

$$\mu_i(\underline{m},\Omega) = \begin{cases} 0, & \text{if } i < \text{ht } \underline{m}. \\ \neq 0, & \text{if } i = \text{ht } \underline{m}. \end{cases}$$

c) [Foxby]. For each maximal ideal $\underline{m}$,

$$\text{id}_{A_{\underline{m}}} \Omega_{\underline{m}} < \infty \quad \text{and} \quad \text{Ext}^i_A(\Omega,\Omega) = \begin{cases} \text{projective, if} & i = 0. \\ 0, \text{ if} & i > 0. \end{cases}$$

d) [Sharp]. For all finitely generated A-modules M,

$$\text{Ext}^i_A(\text{Ext}^j_A(M,\Omega),\Omega) = 0, \quad \text{for} \quad i < j.$$

e) [Sharp]. For each maximal ideal $\underline{m}$,

$$\text{id}_{A_{\underline{m}}} \Omega_{\underline{m}} = \text{depth}_{A_{\underline{m}}} \Omega_{\underline{m}}.$$

[Sharp]. If any one of the above (equivalent) conditions hold, then A is Cohen-Macaulay. QED.

In [12] Bass conjectured that, for a local ring A, there exist nonzero finitely generated A-modules M of finite injective dimension only if A is Cohen-Macaulay. This conjecture has been solved in the affirmative by Peskine and Szpiro [57] in the case of geometric local rings. As pointed out by Sharp [66], Theorem 5.10 provides an affirmative answer when it is assumed that $\text{id}_A M = \text{depth}_A M$.

Having sketched the basic properties of Gorenstein modules in 5.6 - 5.10, we should now like to continue our own investigation of Gorenstein modules with respect to the following two questions:

I. What is the structure of a general Gorenstein module over a local ring? In particular, are they always direct sums of canonical modules?

II. Does every Cohen-Macaulay local ring have a Gorenstein module?

Before beginning this investigation, we need an elementary lemma and a definition. The proof of the lemma is standard and we omit it.

Lemma 5.11. Suppose that A is a local ring with maximal ideal

$\underline{m}$ *and suppose that* Ω *is a Gorenstein* A-*module*.

a) [Sharp; 67] [Foxby; 23]. *If* $\mathfrak{X}$ *is an* A-*sequence, then* $\mathfrak{X}$ *is also an* Ω-*sequence, and* $\Omega/\mathfrak{X}\Omega$ *is a Gorenstein* $A/\mathfrak{X}A$-*module*.

b) *If* $\mathfrak{X}$ *is a maximal* A-*sequence, then* $\Omega/\mathfrak{X}\Omega \cong E^n$, *where* E *is the* $A/\mathfrak{X}A$-*injective envelope of the residue field* $k = A/\underline{m}$. *Moreover* $n = \mu_d(\underline{m},\Omega) = \dim_k \operatorname{Ext}_A^d(k,\Omega)$, *where* $d = \dim A$. *This invariant of* Ω *is called its* ***rank***. QED.

Our next proposition gives a partial answer to Question I (above).

Proposition 5.12. *Suppose* A *is a local ring and* M *and* N *are Gorenstein* A-*modules of the same rank. Then* $M \cong N$.

Proof. The proof goes by induction on $\dim A = \operatorname{depth} A$ (Recall that A is necessarily Cohen-Macaulay (Theorem 5.10)). If $\operatorname{depth} A = 0$, the conclusion follows immediately from Lemma 5.11(b). Now suppose $\operatorname{depth} A = n \geq 1$ and let x be a regular nonunit in A. By Lemma 5.11(a) x is necessarily regular on both M and N. Again by Lemma 5.11, it follows that M/xM and N/xN are Gorenstein A/xA-modules of the same rank; Hence, $M/xM \cong N/xN$, by our induction hypothesis. Therefore, there is an epimorphism $h: M \longrightarrow N/xN$ with $\ker h = xM$. Since necessarily $\operatorname{Ext}_A^1(M,N) = 0$, by [Kaplansky (Theorem 217); 36], we obtain a commutative diagram

$$\begin{array}{ccccccccc} & & & & & & M & & \\ & & & & & \overset{f}{\swarrow} & \downarrow h & & \\ 0 & \longrightarrow & N & \xrightarrow{x} & N & \longrightarrow & N/xN & \longrightarrow & 0. \end{array}$$

An application of Nakayama's Lemma establishes the fact that f is necessarily an epimorphism. Hence, we have an exact sequence $0 \longrightarrow K \longrightarrow M \xrightarrow{f} N \longrightarrow 0$, where $K \subseteq xM$ and $\operatorname{id}_A K < \infty$. Again by [Kaplansky (Theorem 21.7); 36], we have that $\operatorname{Ext}_A^1(N,K) = 0$ and, hence that the above sequence splits. But $K \subseteq xM$ implies that $K = 0$ and, thus, that f is an isomorphism. QED.

Corollary 5.13. (Sharp [69]). *If* A *is a local ring having a canonical module* Ω, *then the Gorenstein* A-*modules* (up to isomorphism) *are precisely the modules* Ω^n, *where* n *is a positive integer*. QED.

In view of Corollary 5.9, we observe that Corollary 5.13 applies to all Cohen-Macaulay, complete local rings. This result was first noted by Sharp in [65]. That Corollary 5.13 does not generalize to commutative Noetherian rings having connected prime spectrum was also noted by Sharp in [69].

With regard to the proof of Proposition 5.12, it was first observed by Foxby [23] that $\mathrm{Ext}^i_A(M,N) = 0$ for $i > 0$, whenever M is a Gorenstein A-module and N is a finitely generated A-module of finite injective dimension.

In general, for a local ring A, we do not know if the existence of a Gorenstein A-module implies the existence of a canonical module (When this is the case, Corollary 5.13 applies). However, we are able to show that, in this situation, some finite, faithfully flat A-algebra posseses a canonical module.

We refer the reader to Auslander and Goldman [7] for terminology concerning central separable algebras, (or central Azumaya algebras), splitting rings and the Brauer group of a commutative ring (denoted $B(A)$, for A a commutative ring).

Theorem 5.14. *Let* A *be a local ring with maximal ideal* $\underline{m}$. *Suppose* Ω *is a Gorenstein* A-*module of rank* n *and let* $\Lambda = \mathrm{End}_A \Omega$.

1) *Then* Λ *is a central separable* A-*algebra*; *Hence*, Λ *represents an element in* $B(A)$.

2) *If* $\mathfrak{X}$ *is any maximal* A-*sequence, then* $\Lambda/\mathfrak{X}\Lambda \cong M_n(A/\mathfrak{X}A)$ ($n \times n$ matrices over $A/\mathfrak{X}A$). *Hence*, Λ *represents an element in the kernel of the maps*

$$B(A) \longrightarrow B(A/\mathfrak{X}A) \quad \text{and} \quad B(A) \longrightarrow B(A/\underline{m}).$$

3) *If any primitive idempotent of* $\Lambda/\underline{m}\Lambda$ *lifts to* Λ, *then* A *has a canonical module* (Hence, Corollary 5.13 applies).

4) *There is a finite commutative* A-*algebra* S *which is free as an* A-*module and which splits* Λ^{op}. *Moreover, if* Ω *is indecomposable*, S *can be chosen so that* Spec S *is connected and* S *has a canonical module*.

Proof. Let $\mathfrak{X}$ be a maximal A-sequence. Then $A/\mathfrak{X}A$ is an artin local ring (Recall that A is Cohen-Macaulay from Theorem 5.10.) and $\mathfrak{X}$ is a maximal Ω-sequence with $\Omega/\mathfrak{X}\Omega \cong E^n$, where E is the $A/\mathfrak{X}A$-injective envelope of $A/\underline{m}$ (Lemma 5.11). By Lemma 5.3 (iii),

$$\Lambda/\mathfrak{X}\Lambda \cong \mathrm{End}_{A/\mathfrak{X}A}(\Omega/\mathfrak{X}\Omega) = \mathrm{End}_{A/\mathfrak{X}A}(E^n) \cong M_n(A/\mathfrak{X}A)$$

since $\mathrm{Hom}_{A/\mathfrak{X}A}(E,E) = A/\mathfrak{X}A$. From Theorem 5.10(C), it is straightforward that the natural map $A \longrightarrow Z(\Lambda)$ (= center of Λ) is a monomorphism and, from the discussion above, it is easily observed that the induced map $A/\mathfrak{X}A \longrightarrow Z(\Lambda)/\mathfrak{X}Z(\Lambda)$ is an epimorphism. Thus, via Nakayama's Lemma the natural map $A \longrightarrow Z(\Lambda)$ is an isomorphism. Hence, parts 1) and 2) are established.

If a primitive idempotent of $\Lambda/\underline{m}\Lambda$ lifts to Λ, it follows that Ω has a direct summand Ω_o (Hence, Ω_o, is a Gorenstein module.) such that $\mathrm{Hom}_A(\Omega_o,\Omega_o) \cong A$. By Theorem 5.6(4), Ω_o is a canonical A-module.

By [Auslander and Goldman (Theorem 6.3); 7], there is a finite commutative A-algebra S which is free as an A-algebra and which splits Λ^{op}, that is, $S \otimes_A \Lambda = \mathrm{Hom}_S(\Lambda^{op},\Lambda^{op})$. ($\Lambda^{op}$ is projective as an S-module). Moreover, from the proof of the aforementioned theorem, S may be chosen as a maximal commutative subring of Λ^{op}. Hence, if Ω is indecomposable, it is clear that S contains no nontrivial idempotents, that is Spec S is connected. It is also clear that S is a semi-local ring and that Λ^{op} is necessarily free as an S-module. Since S is a finite, faithfully flat A-algebra with rad $S = \underline{m}S$, we have $\mathrm{id}_S(S\otimes_A\Omega) < \infty$,

$$\mathrm{Hom}_S(S \otimes_A \Omega,\ S \otimes_A \Omega) \cong S \otimes_A \mathrm{Hom}_A(\Omega,\Omega)$$

$$= S \otimes_A \Lambda \cong \mathrm{Hom}_S(\Lambda^{op},\Lambda^{op}) \quad \text{(projective S-module)}$$

and $\mathrm{Ext}^i_S(S \otimes_A \Omega,\ S \otimes_A \Omega) \cong S \otimes_A \mathrm{Ext}^i_A(\Omega,\Omega) = 0$ if $i > 0$. By Theorem 5.10(c), $S \otimes_A \Omega$ is a Gorenstein S-module. Finally, since

$$\mathrm{End}_S(S \otimes_A \Omega) = \mathrm{Hom}_S(\Lambda^{op},\Lambda^{op}) \cong M_r(S),$$

it follows (as above) that $S \otimes_A \Omega$ has a direct summand which is a canonical S-module. QED.

For the definition and properties of Hensel local rings, we

refer the reader to Nagata [51] and Raynaud [60].

Corollary 5.15. *Let A be a Hensel local ring. Then A has a Gorenstein module if and only if A has a canonical module.* (Hence, Corollary 5.13 applies to Hensel local rings which have Gorenstein modules.)

Proof. Since the sufficiency is obvious, we suppose A is a Hensel local ring which has a Gorenstein module Ω. We may suppose Ω to be indecomposable. By Theorem 5.14(1), (2), $\Lambda = \mathrm{End}_A \Omega$ is a central separable A-algebra such that $\Lambda/\underline{m}\Lambda \cong M_n(A/\underline{m})$, where $n = \mathrm{rank}\ \Omega$. Since A is Hensel, idempotents of $\Lambda/m\Lambda$ lift to idempotents of Λ [See Azumaya, 8, 9]. Thus, necessarily $n = 1$ and Ω is a canonical module. QED.

Our next result generalizes Corollary 5.9.

Theorem 5.16. *A local ring A has a Gorenstein module if and only if A is Cohen-Macaulay and some finite, commutative, faithfully flat A-algebra B with connected prime spectrum is the homomorphic image of a Gorenstein ring of finite Krull dimension.*

Proof. The necessity is a consequence of Theorem 5.14(4) and Corollary 5.9.

We now suppose that A is a Cohen-Macaulay local ring and that B is a finite, commutative, faithfully flat A-algebra with connected prime spectrum which is the homomorphic image of a Gorenstein ring of finite Krull dimension. It is straightforward to check that B is also Cohen-Macaulay and, hence by Corollary 5.9, that B has a canonical module Ω. Furthermore, one may easily verify that (as an A-module) $\mathrm{depth}_A \Omega = \mathrm{depth}\ A = \mathrm{id}_A \Omega$. Thus, as an A-module, Ω is a Gorenstein module (Theorem 5.10(e)). QED.

Let A be a local ring with maximal ideal $\underline{m}$ and residue field k. If $f \in A[x]$, then $\bar{f} \in k[x]$ denotes the polynomial obtained by reducing the coefficients of f modulo $\underline{m}$. The Henselization of A will be denoted by A^h. A standard étale neighborhood of A is a local ring of the form:

$$(A[x]/f)_{\underline{p}} ,$$

where f is a monic polynomial in $A[x]$ such that $f'(0)$ is a unit in A (f' is the derivative of f) and where $\underline{p}$ is a prime ideal which corresponds to the kernel of the homomorphism $g \longmapsto \overline{g}(y)$, $y \in k$ a simple root of $\overline{f}(x) \in k[x]$.

Theorem 5.17. Let A be a local ring.

1) If A has a Gorenstein module, then A^h is a homomorphic image of a Gorenstein local ring (that is A^h has a canonical module).

2) If A^h has a canonical module, then some standard étale neighborhood B of A also has a canonical module.

Proof. 1) Let G be a Gorenstein A-module. Then $\mathrm{depth}_A G = \mathrm{id}_A G$ (Theorem 5.10(e)). From the properties of A^h (See Nagata [51] or Raynaud [60]), it is easily seen that

$$\mathrm{depth}_{A^h}(A^h \otimes_A G) = \mathrm{id}_{A^h}(A^h \otimes_A G) = \mathrm{depth}\ A^h.$$

Again by Theorem 5.10(e), we have that $A^h \otimes_A G$ is a Gorenstein A^h-module. By Corollary 5.15, A^h necessarily has a canonical module and, by Corollary 5.9, A^h is the homomorphic image of a Gorenstein local ring.

2) We now suppose that A^h has a canonical module. Hence there is an exact sequence

$$(A^h)^r \xrightarrow{\varphi} (A^h)^s \longrightarrow \Omega \longrightarrow 0,$$

where the homomorphism φ is given by an $s \times r$ matrix with entries in A^h. Since A^h is a filtered union of standard étale neighborhoods [Nagata (43.9); 51], there is a standard étale neighborhood B of A which contains the entries of the matrix which represents φ. It follows that there is a B-module M such that $\Omega \cong A^h \otimes_B M$. Since, of course, $A^h = B^h$, it is elementary to show that

$$\mathrm{id}_B M < \infty, \quad \mathrm{Hom}_B(M,M) \cong B \quad \text{and}$$

$$\mathrm{Ext}^i_B(M,M) = 0 \quad \text{for} \quad i > 0.$$

(The module Ω has these properties over A^h, by Theorem 5.6(4).) Thus, by Theorem 5.6(4), M is a canonical B-module.

The missing link in the chain of ideas discussed in Theorem 5.17 can be stated in the following fashion: Suppose A is a local ring and suppose B is a faithfully flat, local A-algebra which is essentially of finite type over A. If B has a Gorenstein module, must A necessarily have a Gorenstein module?

With respect to question I concerning the structure of Gorenstein modules over local rings, we now have the following partial result (Theorem 5.17 and Corollary 5.13): If A is a local ring having a Gorenstein module G, then some standard étale neighborhood B of A has a canonical module Ω and $B \otimes_A G \cong \Omega^n$ (n depends on G).

We shall now provide a negative answer to question II, that is, we give an example due to Ferrand and Raynaud [18] of a Cohen-Macaulay local ring which does not have a Gorenstein module. Hence, (Theorem 5.17(1)), there is a Hensel local ring which is not the homomorphic image of a Gorenstein ring (contrary to the situation for complete local rings).

Example 5.18. According to Ferrand and Raynaud [18], there is a 1-dimensional Cohen-Macaulay local domain A which possesses the following properties:

a) The integral closure of A is the Hensel local ring $\mathbb{C}\{x\}$ (the ring of convergent power series over the complex number field).

b) The natural map $\pi : \hat{A}$ (completion of A) $\longrightarrow \mathbb{C}[[x]]$ is surjective with $I = \ker \pi$ nilpotent of index 2. In fact $\hat{A} \cong \mathbb{C}[[x]] \propto I$.

c) The formal fiber $\hat{A} \otimes_A Q$ of A at zero is not a Gorenstein ring (Q denotes the field of quotients of A).

By [Raynaud (Corollary 1, p. 99); 60], the henselization A^h is an integral domain. Hence, the induced map $A^h \longrightarrow \mathbb{C}\{x\}$ is monic. Thus, we may identify A^h in $\mathbb{C}\{x\}$ so that $A \subseteq A^h \subsetneq \mathbb{C}\{x\} \subseteq Q$. Therefore,

$$A^h \otimes_{A^h} Q = \hat{A} \otimes_{A^h} Q \cong \hat{A} \otimes_A Q$$

is not a Gorenstein ring. From Hartshorne [Proposition 10.1; 33], we conclude that A^h does not have a dualizing complex and thus that A^h cannot have a canonical module (the injective resolution of a canonical module is necessarily a dualizing complex in the sense of Hartshorne [33]). It follows, by Corollary 5.9, that A^h is not a homomorphic image of a Gorenstein ring and, by Theorem 5.17(1), that A does not have a Gorenstein module. Actually, in this situation, it is easily established that the only standard étale neighborhood of A is A itself and, thus that $A = A^h$.

Section 6. Dominant dimension of finite algebras

Throughout this section k will denote a commutative artin ring. A k-algebra Λ will be called finite if Λ is finitely generated as a k-module. Such an algebra Λ is necessarily an artin ring. It is known that every finite k-algebra Λ is the homomorphic image of a quasi-Frobenius k-algebra Γ (See Müller [50] or Corollary 4.33(6)). However there is essentially no connection between the homological properties of Λ and those of Γ. In this section of our paper, we establish a method of constructing finite k-algebras Λ_n having dominant dimension $\geq n$ such that $\Lambda \cong \Lambda_n/I_n$, where I_n is a right direct summand of Λ_n (See Theorem 6.2). Thus, we obtain a close connection between the homological properties of Λ and each Λ_n. This construction also enables us to obtain results on the category of finitely generated reflexive modules over a finite k-algebra of dominant dimension ≥ 2 which is right T-stable (See definitions below).

Several papers appear in the literature which bear on our work in this section. Among these are papers of Kato [37, 38, 39], Müller [48, 49, 50], Morita [45, 46, 47] and Tachikawa [70]. Our point of view will be taken partially from the preceding papers and especially from the papers of Gabriel [26] and Roos [64, 65].

Recall from Section 3 that a ring R is n-Gorenstein if, in a minimal (left or right) injective resolution $0 \to R \to E_0 \to E_1 \to \cdots$ of R, flat $\dim_R E_i \leq i$ for $i < n$ (See Theorem 3.7). We say that R is of <u>dominant dimension</u> $\geq n$ (abbreviated dom. dim $R \geq n$) if the stronger property flat $\dim_R E_i = 0$ prevails for $i < n$. A ring R will be called right T-<u>stable</u> if each indecomposable injective in $\underline{\underline{\mathrm{Mod}}}_R$ is either 1-torsion (See Section 3) or torsion free (that is to say, each finitely generated submodule is torsionless in the sense of Bass [10]). If the "1-torsion" condition in the preceding statement is replaced by "torsion" (i.e., no nonzero homomorphism into R), we say that R is <u>right weakly</u> T-<u>stable</u>. If R is right Noetherian and right T-stable, it is easily observed that the Serre subcategory $\mathcal{T}_1$ of 1-torsion modules in $\underline{\underline{Mod}}_R$ is stable in the sense of Roos [64], that is $\mathcal{T}_1$ is closed with respect to injective envelopes. It is also elementary to see that every integral domain is T-stable and every right Noetherian, right hereditary ring is right weakly T-stable.

The results which follow rely heavily upon the earlier results of Sections 1, 3, 4. If Λ is a finite k-algebra, then Λ^d denotes the injective cogenerator $\mathrm{Hom}_k(\Lambda, E)$ considered either as a left or right module, where E is the injective k-module which gives a perfect contravariant duality between injective and projective k-modules. In particular, $\mathrm{End}_\Lambda \Lambda^d \cong \Lambda^{op}$.

Proposition 6.1. *Let Λ be a finite k-algebra, let M be a finitely generated injective module in $\underline{\underline{Mod}}_\Lambda$ and let $\Sigma = \mathrm{End}\, M$. Further, let Γ be the (necessarily finite) k-algebra*

$$\begin{pmatrix} \Lambda & 0 \\ M & \Sigma \end{pmatrix}.$$

Then

(a) $P = \begin{pmatrix} 0 & 0 \\ M & \Sigma \end{pmatrix}$ *is a right projective-injective Γ-module.*

(b) *If $M = \Lambda^d$, then Γ is 1-Gorenstein [Müller; 49] and is both right and left T-stable.*

(c) *Suppose dom. dim $\Lambda \geq m$, Λ is weakly right T-stable and further suppose M is the (right) maximal torsion direct summand of Λ^d. Then dom. dim $\Gamma \geq m + 1$ (dom. dim $\Gamma \geq 2$ if $m = 0$) and Γ is right T-stable.*

(d) *For M as in either (b) or (c), $Q = \begin{pmatrix} \Lambda & 0 \\ 0 & 0 \end{pmatrix}$ is the largest projective right summand of Γ such that $Q/(\mathrm{rad}\, \Gamma)Q$ is torsionfree.*

(e) *For M as in (b) or (c),*

$$\mathrm{gl.\,dim}\, \Lambda \leq \mathrm{gl.\,dim}\, \Gamma \leq 1 + 2\, \mathrm{gl.\,dim}\, \Lambda.$$

Proof. Statement (a) is a direct consequence of the classification of injectives and projectives for the abelian category

$$\underline{\underline{Map}}(F\, \underline{\underline{Mod}}_\Sigma, \underline{\underline{Mod}}_\Lambda) \cong \underline{\underline{Map}}(\underline{\underline{Mod}}_\Sigma, G\, \underline{\underline{Mod}}_\Lambda) \cong \underline{\underline{Mod}}_\Gamma$$

(See Section 1), where $F = - \otimes_{\Sigma} M$ and $G = \mathrm{Hom}_{\Lambda}(M,-)$. In particular, identifying right Γ-modules with objects in $\underline{\underline{\mathrm{Map}}}(F\ \underline{\underline{\mathrm{Mod}}}_{\Sigma}, \underline{\underline{\mathrm{Mod}}}_{\Lambda})$, the maps (See notation in Section 1 and 4)

$$(*) \qquad \begin{array}{c}(0,FI)\\ \downarrow 0 \\ (I,0)\end{array} \qquad\qquad (**) \qquad \begin{array}{c}(0,FGE)\\ \downarrow (0,\eta_E) \\ (GE,E)\end{array}$$

represent injective right Γ-modules, where I is injective in $\underline{\underline{\mathrm{Mod}}}_{\Sigma}$, E is injective in $\underline{\underline{\mathrm{Mod}}}_{\Lambda}$ and $\eta_E : FGE \longrightarrow E$ is the usual natural map $\mathrm{Hom}_{\Lambda}(M,E) \otimes_{\Sigma} M \longrightarrow E$ of evaluation. In our special situation with $E = M$, we see that $(**)$ gives the object

$$\begin{array}{c}(0,M)\\ \downarrow (0,1) \\ (\Sigma,M)\end{array} ,$$

which is naturally equivalent to the right Γ-module $P = \begin{pmatrix} 0 & 0 \\ M & \Sigma \end{pmatrix}$.

(b). To show that Γ is 1-Gorenstein, it suffices to show that Γ (as a right Γ-module) can be embedded in a projective-injective right Γ-module. We already have from part (a) that $P = \begin{pmatrix} 0 & 0 \\ M & \Sigma \end{pmatrix}$ is a projective-injective right Γ-module. Since $\Lambda^d = M$ is an injective cogenerator for $\underline{\underline{\mathrm{Mod}}}_{\Lambda}$, there is an embedding $\Lambda \longrightarrow (\Lambda^d)^n$, for some $n \geq 1$. This embedding gives an embedding of right Γ-modules: $\begin{pmatrix} \Lambda & 0 \\ 0 & 0 \end{pmatrix} \longrightarrow P^n$. Hence, there is a containment (as right Γ-modules): $\Gamma \longrightarrow P^{n+1}$ We remark that this part of (b) as well as the corresponding statement on dominant dimension in part (c) could have been deduced from results in Müller [49]. However, in the interest of selfcontainment, we have chosen to use our own categorical machinery.

We now wish to show that Γ is right T-stable. Firstly, we note that all indecomposable right Γ-injectives are of the form (*) or (**) of the preceding paragraph (Again see Section 1). Secondly, a repetition of the argument used in the preceding paragraph reveals that injectives of type (**) are also projective. Hence, injectives of type (**) are torsion free. Thirdly, to show that injectives of the form

$$\begin{array}{c}(0,FI)\\ \downarrow 0 \\ (I,0)\end{array}$$

(type (*)) are 1-torsion (these injective modules are right Σ-modules as well as right Γ-modules), it suffices to show any right Γ-homomorphism of Σ into Γ is zero. However, this statement follows from the definition of morphisms in

$$\underline{\underline{\mathrm{Map}}}(F\ \underline{\underline{\mathrm{Mod}}}_\Sigma, \underline{\underline{\mathrm{Mod}}}_\Lambda) \cong \underline{\underline{\mathrm{Mod}}}_\Gamma$$

and also from the fact that Λ^d is a faithful left

$$\Sigma = \mathrm{End}_\Lambda\ \Lambda^d \cong \Lambda^{op}\text{-module}.$$

Since Λ^d is left Λ^{op}-injective and since $\mathrm{End}_{\Lambda^{op}}(\Lambda^d) \cong \Lambda$, it follows similarly that Γ is a left T-stable. The above argument also shows that $P/(\mathrm{rad}\ \Gamma)P \cong \Sigma/(\mathrm{rad}\ \Gamma)\Sigma$ is 1-torsion as a right Γ-module. Since Λ^d is an injective cogenerator for Mod_Λ, it follows from the definition of Γ that every right nonzero Λ-module has a non-zero homomorphism into Γ. Thus, $Q = \begin{pmatrix}\Lambda & 0\\ 0 & 0\end{pmatrix}$ is the largest projective right summand of Γ such that $Q/(\mathrm{rad}\ \Gamma)Q$ is torsion free. This proves part (d) in case $M = \Lambda^d$.

(c). Since Λ is a finite k-algebra and weakly right T-stable it is easily seen that $\Lambda^d = M \oplus N$, where $\mathrm{Hom}_\Lambda(M,\Lambda) = 0$ and where N is a projective-injective right Λ-module. To establish that Γ is right T-stable, one only need modify slightly the corresponding argument given in part (b). The same statement applies to the verification of part (d) in this case. Now let

$$0 \longrightarrow \Lambda \longrightarrow E_0 \longrightarrow E_1 \longrightarrow \cdots$$

be a right minimal injective resolution of Λ, where by assumption the E_i's are projective (flat = projective for artin rings), for $i < m$. Since $\mathrm{Hom}_\Lambda(M,\Lambda) = \mathrm{Hom}_\Lambda(M,E_i) = 0$ for $i < m$, we obtain an exact Γ-injective complex of the form (notation as in Section 1)

$$\begin{array}{ccccccccc}
(0,0) & \longrightarrow & (0,0) & \longrightarrow & \cdots & \longrightarrow & (0,0) & \longrightarrow & (0,FGE_m)\\
\downarrow & & \downarrow & & & & \downarrow & & \downarrow (0,\eta)\\
(0,E_1) & \longrightarrow & (0,E_2) & \longrightarrow & \cdots & \longrightarrow & (0,E_{m-1}) & \longrightarrow & (GE_m,E_m)
\end{array}$$

where the injectives are of type (**) (See part (a) of this proof). Moreover, the zero$\underline{\mathrm{th}}$ homology of this complex is

$$\begin{array}{c}(0,0)\\ \downarrow \\ (0,\Lambda)\end{array}$$

which is the $\underline{\underline{\text{Map}}}(\ F\ \underline{\underline{\text{Mod}}}_\Sigma, \underline{\underline{\text{Mod}}}_\Lambda)$ equivalent form of the right Γ-module $Q = \begin{pmatrix}\Lambda & 0\\ 0 & 0\end{pmatrix}$. In addition, each of the injectives in the complex are also projective as right Γ-modules. Since $P = \begin{pmatrix}0 & 0\\ M & \Sigma\end{pmatrix}$ is a projective-injective right Γ-module (part (a)), it now follows that dom. dim $\Gamma \geq m+1$. The special case for $m = 0$ follows along the same lines.

(d). The proof of this statement is contained in the proofs of (b) and (c) as indicated above.

(e). If $M = \Lambda^d$, then Corollary 4.21 gives

$$\text{gl. dim}\ \Lambda \leq \text{gl. dim}\ \Gamma \leq 1 + \text{gl. dim}\ \Lambda + \text{gl. dim}\ \Lambda^{op}.$$

However, gl. dim Λ = gl. dim Λ^{op}. In the case of part (c), we have

$$\text{gl. dim}\ \Lambda \leq \text{gl. dim}\ \Gamma \leq 1 + \text{gl. dim}\ \Lambda + \text{gl. dim}\ \Sigma.$$

Since $\Lambda^d = N \oplus M$, where N is right Λ-projective and

$$\text{Hom}_\Lambda(M,\Lambda) = 0,$$

it follows that

$$\Lambda^{op} \cong \text{End}_\Lambda\ \Lambda^d = \text{End}\ (N \oplus M) \cong$$

$$\begin{pmatrix}\text{End}\ N & \text{Hom}(M,N)\\ \text{Hom}(N,M) & \text{End}\ M\end{pmatrix} \cong \begin{pmatrix}\text{End}\ N & 0\\ \text{Hom}(N,M) & \Sigma\end{pmatrix}.$$

By Corollary 4.21,

$$\text{gl. dim}\ \Sigma \leq \text{gl. dim}\ \Lambda^{op} = \text{gl. dim}\ \Lambda.$$

Thus, in case either part (b) or part (c) holds, we have the desired inequality concerning the global dimension of Γ. QED.

Among other things, our next theorem shows that finite k-algebras of dominant dimension at least n are not particularly rare.

Theorem 6.2. *Let* Λ *be a finite* k-*algebra*. *For every* $n > 0$ *there is a finite* k-*algebra* Λ_n *and a two sided ideal* I_n *of* Λ_n *satisfying*:

(1). $\Lambda \cong \Lambda_n/I_n$ *and* I_n *is a right direct summand of* Λ_n (hence right projective).

(2). Λ_n *is right* T-*stable*.

(3). dom. dim $\Lambda_n \geq n$.

(4). gl. dim $\Lambda \leq$ gl. dim $\Lambda_n \leq 2^{n-1} + (2^{n-1}+1)$gl. dim Λ.

Proof. Let $\Lambda_1 = \begin{pmatrix} \Lambda & 0 \\ {}_\Lambda{}^d{}_\Lambda & \Lambda^{op} \end{pmatrix}$. Then Statements (1), (2), (3) and (4) hold for Λ_1, by Proposition 6.1. Assuming Λ_i has been constructed, for $1 \leq i \leq n-1$, so that (1) - (4) hold, define

$$\Lambda_n = \begin{pmatrix} \Lambda_{n-1} & 0 \\ t(\Lambda^d_{n-1}) & \Sigma_{n-1} \end{pmatrix},$$

where $t(\Lambda^d_{n-1})$ is the maximal right torsion summand of Λ^d_{n-1} and $\Sigma_{n-1} = \mathrm{End}_{\Lambda_{n-1}}(t(\Lambda^d_{n-1}))$. An elementary induction argument together with Proposition 6.1 shows that (1) - (4) hold for Λ_n. QED.

Example 6.3. Starting with $\Lambda = k$, k a field, and applying the construction of Λ_n as described in Theorem 6.2, we get the following well-known class of rings:

$$\Lambda_n = T_{n+1}(k)/\mathrm{rad}\ T_{n+1}(k)^2 = \begin{pmatrix} k & & & 0 \\ k & k & & \\ & k & \ddots & \\ 0 & & k & k \end{pmatrix}.$$

It is known that gl. dim $\Lambda_n = n$. From Theorem 6.2, we may conclude dom. dim $\Lambda_n = n$, since necessarily dom. dim $\Lambda_n \leq$ gl. dim Λ_n. Hence, this gives an example where the increase of the dominant dimension is exactly one at each stage of the construction in Theorem 6.2.

Now let Γ be a finite k-algebra with dom. dim $\Gamma \geq 2$, let $\mathcal{T}_1$ be the Serre subcategory of 1-torsion modules in $\underline{\underline{\mathrm{Mod}}}_\Gamma$ and let $\underline{\underline{R}}_\Gamma$

be the full subcategory of finitely generated (= coherent) reflexive modules in $\underline{\underline{\mathrm{Mod}}}_\Gamma$. It is easy to show (See Morita [46,47]) that $\underline{\underline{R}}_\Gamma$ consists precisely of the finitely generated $\mathcal{T}_1$-closed objects of $\underline{\underline{\mathrm{Mod}}}_\Gamma$ (in the sense of Gabrial [26]). Further, let P be the direct sum of the projective indecomposable modules X in $\underline{\underline{\mathrm{Mod}}}_\Gamma$ such that $X/(\mathrm{rad}\ \Gamma)X \notin \mathcal{T}_1$. Then results of Morita [46,47] and Gabriel [26] show the following categories are equivalent [$\underline{\underline{\mathrm{Coh}}}$ ($\underline{\underline{\mathrm{Mod}}}_R$)= finitely generated right R-modules when R is finite k-algebra]:

(a) $\underline{\underline{\mathrm{Coh}}}(\underline{\underline{\mathrm{Mod}}}_\Gamma/\mathcal{T}_1)$

(b) $\underline{\underline{\mathrm{Coh}}}(\underline{\underline{\mathrm{Mod}}}_{\mathrm{End}_\Gamma P})$

(c) $\underline{\underline{R}}_\Gamma$

In particular, $\underline{\underline{R}}_\Gamma$ is an abelian category. If Λ is a weakly right T-stable finite k-algebra and if Λ is an in Proposition 6.1(c), then the projective P (as in part (b) above) for Γ is of the form $\begin{pmatrix}\Lambda & 0\\ 0 & 0\end{pmatrix}$ (See Proposition 6.1 (d)). Hence $\underline{\underline{R}}_\Gamma$ is equivalent to $\underline{\underline{\mathrm{Coh}}}(\underline{\underline{\mathrm{Mod}}}_{\mathrm{End}_\Gamma P})$ $\cong \underline{\underline{\mathrm{Coh}}}(\underline{\underline{\mathrm{Mod}}}_\Lambda)$. This discussion together with Proposition 6.1 yields our next result.

Theorem 6.4. Let Λ be a weakly right T-stable, finite k-algebra. Then $\underline{\underline{\mathrm{Coh}}}(\underline{\underline{\mathrm{Mod}}}_\Lambda)$ is equivalent to the category of finitely generated reflexive right modules over some right T-stable, finite k-algebra Γ with dom. dim $\Gamma \geq 2$ and

$$\text{gl. dim}\,\Lambda \leq \text{gl. dim}\,\Gamma \leq 1 + 2\ \text{gl. dim}\,\Lambda.$$

QED.

Corollary 6.5. If Λ is an hereditary finite k-algebra, then $\underline{\underline{\mathrm{Coh}}}(\underline{\underline{\mathrm{Mod}}}_\Lambda)$ can be realized as the category of finitely generated reflexive right modules over some finite k-algebra Γ with

$$\text{dom. dim}\,\Gamma \geq 2 \quad\text{and}\quad \text{gl. dim}\,\Gamma \leq 3.$$

Proof. As noted in the introduction of this section, every hereditary finite k-algebra is weakly right T-stable. QED.

Corollary 6.6. Let Λ be an arbitrary finite k-algebra and let

$\Phi = \begin{pmatrix} \Lambda & 0 \\ \Lambda^d & \Lambda^{op} \end{pmatrix}$. Then $\underline{\underline{Coh}}(\underline{\underline{Mod}}_\Phi)$ can be realized as the category of finitely generated reflexive right modules over some finite k-algebra Γ with dom. dim $\Gamma \geq 2$ and

$$\text{gl. dim } \Lambda \leq \text{gl. dim } \Gamma \leq 2 + 3 \text{ gl. dim } \Lambda.$$

Proof. This statement follows from Theorem 6.4 and Proposition 6.1. QED.

As a sort of converse to Theorem 6.4, we prove the following result.

Theorem 6.7. Let Γ be a right T-stable, finite k-algebra with dom. dim $\Gamma \geq 2$. Then the abelian category $\underline{\underline{R}}_\Gamma$ of finitely generated reflexive right Γ-modules is equivalent to $\underline{\underline{Coh}}(\underline{\underline{Mod}}_\Lambda)$ for some weakly right T-stable, finite k-algebra Λ.

Proof. The proof proceeds along the lines of the discussion preceding Theorem 6.4. So let P be the direct sum of the projective indecomposable modules X in $\underline{\underline{Mod}}_\Gamma$ such that

$$X/(\text{rad } \Gamma)X \notin \mathcal{T}_1 \quad (= \text{1-torsion objects in } \underline{\underline{Mod}}_\Gamma),$$

let $\Lambda = \text{End}_\Gamma P$ and let $\underline{\underline{Q}}$ denote the quotient category $\text{Mod}_\Gamma/\mathcal{T}_1$. As noted in the discussion previous to Theorem 6.4, the categories $\underline{\underline{Coh}}\ \underline{\underline{Q}}$, $\underline{\underline{Coh}}(\underline{\underline{Mod}}_\Lambda)$ and $\underline{\underline{R}}_\Gamma$ are equivalent. (In fact, $\underline{\underline{Q}}$ and $\underline{\underline{Mod}}_\Lambda$ are equivalent categories.) Thus, our proof that Λ is weakly right T-stable will be constructed in the category $\underline{\underline{Q}}$. Recall that the objects of $\underline{\underline{Q}}$ are the same as those of $\underline{\underline{Mod}}_\Gamma$; However,

$$\text{Hom}_{\underline{\underline{Q}}}(A,B) = \varinjlim \text{Hom}(A', B/B'),$$

where $A/A', B' \in \mathcal{T}_1$ (See Gabriel [26]). Since the objects in $\underline{\underline{Coh}}(\underline{\underline{Mod}}_\Gamma)$ are artinian, it easily follows, for $A,B \in \underline{\underline{Coh}}(\underline{\underline{Mod}}_\Gamma)$, that

$$\text{Hom}_{\underline{\underline{Q}}}(A,B) = \text{Hom}_\Gamma(\bar{A}, B/tB),$$

where $\bar{A}$ is the unique maximal submodule of A such that $A/\bar{A} \in \mathcal{T}_1$ and tB is the unique maximal submodule of B which is also in $\mathcal{T}_1$.

Since Γ is right T-stable and dom. dim $\Gamma \geq 2$, we have that $\Gamma^d = M \oplus N$, where $N \in \mathcal{T}_1$ and $M \neq 0$ is a projective-injective module in $\underline{\underline{\mathrm{Mod}}}_\Gamma$. In addition, $M = I \oplus J$, where $I/(\mathrm{rad}\ \Gamma)I$ is torsionfree and $J/(\mathrm{rad}\ \Gamma)J \in \mathcal{T}_1$. As an object in $\underline{\underline{Q}}$, $M = I \oplus J$ is an injective-cogenerator. It is also clear from the properties of I and P (See first paragraph) that $I \longrightarrow P^n$, for some $n \geq 1$. Hence I is torsion-free in $\underline{\underline{Q}}$. It remains to show that J is a torsion object in $\underline{\underline{Q}}$, that is, $0 = \mathrm{Hom}_{\underline{\underline{Q}}}(J,P) \cong \mathrm{Hom}_\Gamma(\overline{J},P)$ (Recall P is torsion free, i.e., $tP = 0$). To this end we first observe that $\mathrm{Hom}_\Gamma(T,P/V) = 0$ for all $T \in \mathcal{T}_1$ and $V \subsetneq P$; For a nonzero map in $\mathrm{Hom}_\Gamma(T,P/V)$ would induce a nonzero map $P \longrightarrow E(T)$ (= Injective envelope of T; necessarily a module in $\mathcal{T}_1$) which would contradict the fact that $P/(\mathrm{rad}\ \Gamma)P$ is torsionfree. Secondly, if $P \subset E_o(P)$ denotes the inclusion of P into its injective envelope, we have that $E_o(P)/P$ is torsionfree, since dom. dim $\Gamma \geq 2$.

Now suppose $\varphi : \overline{J} \longrightarrow P$ is a nonzero Γ-map. Then φ extends to a map $\varphi^* : J \longrightarrow E_o(P)$, since $E_o(P)$ is injective. Since $J/\overline{J} \neq 0 \in \mathcal{T}_1$ and since $\mathrm{Hom}_\Gamma(T,\Gamma/V) = 0$, for all $T \in \mathcal{T}_1$ and $V \subsetneq P$, it follows that φ^* induces a nonzero monomorphism $\Psi : J/\overline{J} \longrightarrow E_o(P)/P$. But his contradicts the fact that $E_o(P)/P$ is torsionfree. Thus, $\mathrm{Hom}_{\underline{\underline{Q}}}(J,P) \cong \mathrm{Hom}_\Gamma(\overline{J},P) = 0$. QED.

We refer the reader to Morita [46] and Reiten [62] for related results concerning the category of reflexive modules over Artin rings of dominant dimension ≥ 2.

Section 7. Representation dimension of finite algebras

In this section, as in section 6, we shall be considering finite k-algebras over a commutative Artin ring k.

In [4], M. Auslander has established the following results: Let Λ be a finite k-algebra of finite representation type (that is, a finite number of indecomposable modules up to isomorphism), let M be the direct sum of the distinct indecomposable modules in ${}_{\Lambda}\underline{\underline{Mod}}$ and let $\Gamma = \mathrm{End}_{\Lambda}M$. Then dom. dim $\Gamma \geq 2$ and gl. dim $\Gamma \leq 2$, (in fact gl.dim $\Gamma = 2$, unless Λ is semi-simple). Moreover, ${}_{\Lambda}\underline{\underline{Mod}}$ is equivalent to $\mathrm{End}_{\Gamma}E_{o}\underline{\underline{Mod}}$, where E_o is the injective envelope of Γ in ${}_{\Gamma}\underline{\underline{Mod}}$.

Conversely, if Γ is a finite k-algebra such that gl. dim $\Gamma \leq 2$ and if P is a finitely generated projective-injective module in ${}_{\Gamma}\underline{\underline{Mod}}$, then $\mathrm{End}_{\Gamma}P$ is a finite k-algebra of finite representation type.

The proof of the first half of this result makes use of the fact that the functor category $\underline{\underline{Coh}}[({}_{\Lambda}\underline{\underline{Mod}})^{op},\underline{\underline{Ab}}]$ is equivalent to the category $\underline{\underline{Coh}}({}_{\Gamma}\underline{\underline{Mod}})$ = finitely generated left Γ-modules.

Let Λ be a finite k-algebra and let $\underline{\underline{A}}$ be a full additive subcategory of ${}_{\Lambda}\underline{\underline{Mod}}$ generated by a finite number of indecomposable modules in ${}_{\Lambda}\underline{\underline{Mod}}$ such that $\underline{\underline{A}}$ contains all projective and injective indecomposable modules in ${}_{\Lambda}\underline{\underline{Mod}}$. Then Auslander [4] shows that $\underline{\underline{A}}$ is coherent in the sense of Section 2 and that

$$\text{dom. dim } \underline{\underline{Coh}}[\underline{\underline{A}}^{op},\underline{\underline{Ab}}] \geq 2.$$

(Actually, Morita [46] shows $\underline{\underline{A}}$ must contain the indecomposable projectives and indecomposable injectives in order for

$$\text{dom. dim } \underline{\underline{Coh}}[\underline{\underline{A}}^{op},\underline{\underline{Ab}}] \geq 2).$$

Define the left representation dimension of Λ by

$$\text{left rep. dim } \Lambda = \inf_{\underline{\underline{A}}}\{\text{gl. dim } \underline{\underline{Coh}}[\underline{\underline{A}}^{op},\underline{\underline{Ab}}]\} .$$

If we let A be the direct sum of each of each of the distinct indecomposable modules in $\underline{\underline{A}}$, then Auslander [4] shows

$$\text{gl. dim } \text{End}_{\Lambda} A = \text{gl. dim } \underline{\underline{\text{Coh}}}[\underline{\underline{A}}^{op}, \underline{\underline{\text{Ab}}}].$$

After Morita [46], we call a module A (with the above properties) a coherent generator-cogenerator of ${}_{\Lambda}\text{Mod}$. Thus, left rep. dim Λ = inf{gl. dim $\text{End}_{\Lambda} A$: A is a coherent, generator-cogenerator of ${}_{\Lambda}\underline{\underline{\text{Mod}}}$}. Using the duality of the dualizing injective E for k (See Section 6), it is easily observed that

$$\text{left rep. dim } \Lambda = \text{right rep. dim } \Lambda.$$

Thus, we speak only of the representation dimension of Λ, denoted by rep. dim Λ. The results that follow exhibit several classes of finite k-algebras having finite representation dimension. We note, from the preceding theory due to M. Auslander, if rep. dim $\Lambda \leq 2$, then Λ is of finite representation type.

Theorem 7.1. *If Λ is a weakly right* T-*stable* (See Section 6), *finite* k-*algebra and if* gl. dim $\Lambda < \infty$, *then*

$$\text{rep. dim } \Lambda \leq 1 + 2 \text{ gl. dim } \Lambda.$$

Proof. Since Λ is right weakly T-stable, $\Lambda^d = N \oplus t\Lambda^d$, where N is a right projective-injective Λ-module and $\text{Hom}_{\Lambda}(t\Lambda^d, \Lambda) = 0$. Let $A = \Lambda \oplus t\Lambda^d \in \underline{\underline{\text{Mod}}}_{\Lambda}$. Then clearly A is a coherent, generator-cogenerator of $\underline{\underline{\text{Mod}}}_{\Lambda}$ and

$$\text{End}_{\Lambda} A \cong \begin{pmatrix} \Lambda & 0 \\ t\Lambda^d & \text{End}_{\Lambda} t\Lambda^d \end{pmatrix},$$

since $\text{Hom}_{\Lambda}(t\Lambda^d, \Lambda) = 0$. By Proposition 6.1(e),

$$\text{gl. dim End } A \leq 1 + 2 \text{ gl. dim } \Lambda.$$

It follows by definition that rep. dim $\Lambda \leq 1 + 2$ gl. dim Λ. QED.

Corollary 7.2. *If Λ is an hereditary, finite* k-*algebra, then* rep. dim $\Lambda \leq 3$. *If Λ is not of finite representation type, then* rep. dim $\Lambda = 3$. QED.

If Λ is a finite k-algebra with gl. dim $\Lambda < \infty$, then (in the notation of Theorem 6.2)

$$\text{rep. dim } \Lambda_n \leq 2^n + (2^n + 1) \text{ gl. dim } \Lambda .$$

Now invoking Theorem 7.1 and Theorem 6.2 we see that every finite k-algebra of finite global dimension is a homomorphic image of a finite k-algebra having both finite global dimension and finite representation dimension. We also remark that Auslander [4] has shown: If Λ is a Q.F., finite k-algebra, than

$$\text{rep. dim } \Lambda \leq \text{Loewy length of } \Lambda < \infty.$$

Theorem 7.3. <u>Let</u> Λ <u>be a finite</u> k-<u>algebra and let</u> $T_2(\Lambda)$ <u>denote the ring of</u> 2×2 <u>lower triangular matrices over</u> Λ. <u>Then</u>

$$\text{rep. dim } T_2(\Lambda) \leq 2 + \text{rep. dim } \Lambda.$$

Proof. We may assume that rep. dim $\Lambda = n < \infty$. Hence, there is a coherent, generator-cogenerator A in ${}_\Lambda\underline{\underline{\text{Mod}}}$ such that gl. dim $\text{End}_\Lambda A = n$. We use the notation of Section 1 to describe left modules over $T_2(\Lambda) = \begin{pmatrix} \Lambda & 0 \\ \Lambda & \Lambda \end{pmatrix}$, that is, we think of left $T_2(\Lambda)$-modules as objects in $\underline{\underline{\text{Map}}}(F_\Lambda\underline{\underline{\text{Mod}}}, {}_\Lambda\underline{\underline{\text{Mod}}})$, where $F = \Lambda \otimes_\Lambda -$ = identity functor. With this convention in mind, we let $\bar{A}$ denote the left $T_2(\Lambda)$-module which is the direct sum of the modules

$$\bar{A}_1 = \begin{matrix} (0,0) \\ \downarrow 0 \\ (0,A) \end{matrix}, \quad \bar{A}_2 = \begin{matrix} (0,A) \\ \downarrow (0,1) \\ (A,A) \end{matrix} \quad \text{and} \quad \bar{A}_3 = \begin{matrix} (0,A) \\ \downarrow \\ (A,0) \end{matrix} .$$

The description of projectives and injectives in Section 1 for

$$\underline{\underline{\text{Map}}}(F_\Lambda\underline{\underline{\text{Mod}}}, {}_\Lambda\underline{\underline{\text{Mod}}}) = \underline{\underline{\text{Map}}}({}_\Lambda\underline{\underline{\text{Mod}}}, {}_\Lambda\underline{\underline{\text{Mod}}})$$

shows that

$$\bar{A} = \bar{A}_1 \oplus \bar{A}_2 \oplus \bar{A}_3$$

is also a coherent, generator-cogenerator for

$$\underline{\underline{\text{Map}}}({}_\Lambda\underline{\underline{\text{Mod}}}, {}_\Lambda\underline{\underline{\text{Mod}}}) \cong {}_{T_2(\Lambda)}\underline{\underline{\text{Mod}}}.$$

An examination of morphisms in $\underline{\underline{\text{Map}}}({}_\Lambda\underline{\underline{\text{Mod}}}, {}_\Lambda\underline{\underline{\text{Mod}}})$ shows that

$$\text{Hom}(\bar{A}_2, \bar{A}_1) \cong \text{Hom}(\bar{A}_3, \bar{A}_1) \cong \text{Hom}(\bar{A}_3, \bar{A}) \cong \text{Hom}(\bar{A}, \bar{A}) = 0$$

and otherwise all homomorphism groups give $\mathrm{Hom}_\Lambda(A,A) = \Sigma$. Thus

$$\mathrm{End}_{T_2(\Lambda)}(\bar{A}) = \mathrm{End}_{T_2(\Lambda)}(\bar{A}_1 \oplus \bar{A}_2 \oplus \bar{A}_3)$$

$$\cong \begin{pmatrix} \Sigma & 0 & 0 \\ \Sigma & \Sigma & 0 \\ 0 & \Sigma & \Sigma \end{pmatrix}$$

Let $\Gamma = \begin{pmatrix} \Sigma & 0 \\ \Sigma & \Sigma \end{pmatrix}$. Then $\mathrm{End}_{T_2(\Lambda)}(\bar{A}) \cong \begin{pmatrix} \Gamma & 0 \\ M & \Sigma \end{pmatrix}$, where $\mathrm{pd}_\Sigma M = 0$. Hence. gl. dim $\mathrm{End}_{T_2(\Lambda)}(\bar{A}) \leq \max(\text{gl. dim } \Gamma + 0 + 1, \text{gl. dim } \Sigma) \leq \max(n + 1 + 1, n) = n + 2$ (See Corollary 4.21 (4)). QED.

Let $\mathbb{Z}(p^n)$ denote the ring of integers modulo p^n, where p is a prime in $\mathbb{Z}$. M. Auslander has shown that rep. dim $T_2(\mathbb{Z}(p^2)) = 2$ and J. Janusz further established rep. dim $T_2(Z(p^3)) = 2$ (See Auslander [4] for more details). In any case, we have rep. dim $T_2(\mathbb{Z}(p^n)) \leq 4$ by Theorem 7.3 (Of course rep. dim $\mathbb{Z}(p^n) \leq 2$ for all n). Furthermore, S. Brenner [13] has shown that rep. dim $T_2(\mathbb{Z}(p^n)) = 3$ or 4 for $n \geq 4$, but it is not known which value is correct.

REFERENCES

1. M. Auslander, On the dimension of modules and **algebras** III: Global dimension, Nagoya Math. J., 9, 67-77 (1955).

2. M. Auslander, Coherent functors, Proceedings of the Conference on Categorical Algebra, La Jolla 1965 (Springer-Verlag, 189-231.

3. M. Auslander, Anneaux de Gorenstein et torsion en algébre commutative (texte rédigé par Mangeney, Peskine and Szpiro) Seminaire d' algébre Commutative 1966/67. Paris: Secrétariat Mathématique, 1967.

4. M. Auslander, Representation dimension of Artin algebras, Queen Mary College Math. Notes, London, 1971.

5. M. Auslander and M. Bridger, Stable module theory, Mem. Amer. Math. Soc. No. 94 (1969).

6. M. Auslander and D. Buchsbaum, Homological dimension in local rings. Trans. Amer. Math. Soc., 85, 390-405 (1957).

7. M. Auslander and O. Goldman, The Brauer group of a commutative ring, Trans. Amer. Math. Soc., 97, 367-409 (1960).

8. G. Azumaya, On maximally central algebras, Nagoya Math. J., 2, 119-150 (1951).

9. G. Azumaya, Algebras with Hochschild Dimension ≤ 1, In :Ring Theory, pp. 9-27. New York and London: Academic Press, 1972.

10. H. Bass, Finitistic dimension and a homological generalization of semi-primary rings, Trans, Amer. Math. Soc., 95, 466-488 (1960).

11. H. Bass, Injective dimension in Noetherian rings, Trans. Amer. Math. Soc., 102, 18-29 (1962).

12. H. Bass, On the ubiquity of Gorenstein rings, Math. Z., 82, 8-28 (1963)

13. S. Brenner, Large indecomposable modules over a ring of 2x2 triangular matrices, Bull. London Math. Soc., 3, 333-336 (1971).

14. H. Cartan and S. Eilenberg, Homological Algebra, Princeton University Press, Princeton, 1956.

15. S. Chase, A generalization of the ring of triangular matrices, Nagoya Math. J., 18, 13-25 (1961).

16. I. S. Cohen, On the structure and ideal theory of complete local rings, Trans. Amer. Math. Soc., 59, 54-106 (1946).

17. S. Eilenberg, A. Rosenberg, D. Zelinsky, On the dimension of modules and algebras VIII. Dimension of tensor products, Nagayo Math. J., 12, 71-93 (1957).

18. D. Ferrand and M. Raynaud, Fibres formelles d'un anneau local Noetherien, Ann. Sci. Ec. Norm., (4) 3, 295-311 (1970).

19. K. Fields, On the global dimension of residue rings, Pac. J. Math., 32, 345-349 (1971).

20. R. Fossum, Duality over Gorenstein rings, Math. Scand., 26, 165-176 (1970).

21. R. Fossum and I. Reiten, Commutative n-Gorenstein rings, Math. Scand. 31, 33-48 (1972).

22. R. Fossum, P. Griffith, I. Reiten, Trivial extensions of abelian categories and applications to rings: an expository account. In: Ring Theory, pp. 125-151. New York and London: Academic Press, 1972.

23. H.-B. Foxby, Gorenstein modules and related modules, Math. Scand., 31, 267-284 (1972).

24. H.-B. Foxby, Cohen-Macaulay modules and Gorenstein modules, Math. Scand. 31, 267-284 (1972).

25. P. Freyd, Abelian categories, Harper and Row, New York, 1964.

26. P. Gabriel, Des catégories abéliennes, Bull. Soc. Math. France, 90, 323-448 (1962)

27. A. Grothendieck, Théorèms de dualitè pour les faisceaux algébriques cohérents. Séminaire Bourbaki, May 1957.

28. A. Grothendieck, Local Cohomology, Berlin-Heidelberg-New York. Springer 1967 (Lecture Notes in Mathematics No. 41).

29. A. Grothendieck and J. Dieudonné, Elements de Géométrie algébrique IV (sec. patie). Publ. Math. I. H. E. S. No. 24 (1965).

30. T. Gulliksen, Massey operations and the Poincaré series of certain local rings, J. Algebra 22, 223-232 (1972).

31. M. Harada, Hereditary semiprimary rings and triangular matrix rings, Nogoya Math. J., 27, 463-484 (1966).

32. M. Harada, On special type of hereditary abelian categories, Osaka J. Math., 4, 243-255 (1967).

33. R. Hartshorne, Residues and duality, Berlin-Heidelberg-New York: Springer 1966 (Lecture Notes in Mathematics No. 20).

34. J. Herzog and E. Kunz, Der kanonische Modul eines Cohen-Macaulay Rings, Berlin-Heidelberg-New York: Springer 1971 (Lecture Notes in Mathematics No. 238).

35. A. V. Jategaonkar, A counter-example in ring theory and homological algebra, J. of Algebra, 12, 418-440 (1969).

36. I. Kaplansky, Commutative rings, Allyn and Bacon, Inc., Boston, 1970.

37. T. Kato, Rings of dominant dimension ≥ 1. Proc. Japan Acad., 44, 579-584 (1968).

38. T. Kato, Rings of U-dominant dimension ≥ 1, Tohoku Math. J., 21, 321-327 (1969).

39. T. Kato, Dominant modules, J. of Algebra, 14, 341-349 (1970).

40. E. Matlis, Injective modules over Noetherian rings, Pac. J. Math., 8, 511-528 (1958).

41. B. Mitchell, Theory of Categories, Academic Press, New York and London, 1965.

42. B. Mitchell, On the dimension of objects and categories I, J. of Algebra, 9, 314-340 (1968).

43. B. Mitchell, On the dimension of objects and categories II, J. of Algebra, 9, 341-368 (1968).

44. B. Mitchell, Rings with several objects, Advances in Mathematics, 8, 1-161 (1972).

45. K. Morita, Duality in Q.F.-3 rings, Math. Z., 108, 237-252 (1969).

46. K. Morita, Localizations in categories of modules I, Math. Z., 114, 121-144 (1970).

47. K. Morita, Localizations in categories of modules II, Crelles J., 242, 163-169 (1970).

48. B. J. Müller, The classification of algebras by dominant dimension, Can. J. Math., 20, 398-409 (1968).

49. B. J. Müller, Dominant dimension of semi-primary rings, Crelles J., 232, 173-179 (1968).

50. B. J. Müller, On Morita duality, Can. J. Math., 21, 1338-1347 (1969

51. M. Nagata, Local rings, Wiley and Sons, New York, London, 1962.

52. T. Nakayama, Note on uniserial and generalized uniserial rings, Proc. Imp. Soc. Japan, 16, 285-289 (1940).

53. U. Oberst and H. Röhrl, Flat and coherent functors, J. of Algebra, 14, 91-105 (1970).

54. B. L. Osofsky, Homological dimension and the continuum hypothesis, Trans. Amer. Math. Soc., 132, 217-231 (1968).

55. B. L. Osofsky, Homological dimension and cardinality, Trans. Amer. Soc., 151, 641-649 (1970).

56. I. Palmer and J.-E. Roos, Formules explicites pour la dimension homologique des anneaux de matrices généralixées, Comptes rendus, série A. 273, 1026-1029 (1971)

57. I. Palmer and J.-E. Roos, Explicit formulae for the global homological dimensions of trivial extensions of rings, J. of Algebra 27, 380-413 (1974).

58. C. Peskine and L. Szpiro, Sur la topologie des sous-schémes fermés d'un schéma localement noethérien définis comme support d'un faisceau cohérent localement de dimension projective finie, C. R. Acad. Sci. Paris, 269, 49-51 (1969).

59. D. Quillen, On the (co-) homology of commutative rings, Proc. Symp. Pure Math., 17, 65-87 (1970).

60. M. Raynaud, Anneaux Locaux Henséliens, Berlin-Heidelberg-New York: Springer 1970 (Lecture Notes in Mathematics No. 169).

61. M. Raynaud and L. Gruson, Critéres de platitude et de projectivé, Inventiones Math., 13, 1-89 (1971).

62. I. Rieten, Trivial extensions of Gorenstein rings, thesis, University of Illinois, Urbana, 1971.

63. I. Reiten: The converse to a theorem of Sharp on Gorenstein modules Proc., of A.M.S., 32, 417-420 (1970).

64. J.-E. Roos, Locally Noetherian categories and generalized strictly linearly compact rings. Applications, Category Theory, Homology Theory and their applications II, Lecture Notes in Mathematics 92, Springer-Verlag: Berlin 1969.

65. J.-E. Roos, Coherence of general matrix rings and non-stable extensions of locally Noetherian categories, mimeographed notes, University of Lund.

66. R. Sharp, The Cousin complex for a module over a commutative Noetherian ring, Math. Z., 115, 117-139 (1970).

67. R. Sharp, Gorenstein modules, Math. Z., 115, 117-139 (1970).

68. R. Sharp, On Gorenstein modules over a complete Cohen-Macaulay ring, Oxford Quart. J. Math., 22, 425-434 (1971).

69. R. Sharp, Finitely generated modules of finite injective dimension over certain Cohen-Macaulay rings. Proc. London. Math. Soc. (3) 25, 303-328 (1972).

70. H. Tachikawa, On left QF-3 rings, Pac. J. Math., 32, 255-268 (1970)

71. C. E. Watts, Intrinsic characterization of some additive functors, Proc. Amer. Math. Soc., 11, 5-8 (1960).

Vol. 277: Séminaire Banach. Edité par C. Houzel. VII, 229 pages. 1972. DM 22,-

Vol. 278: H. Jacquet, Automorphic Forms on GL(2) Part II. XIII, 142 pages. 1972. DM 18,-

Vol. 279: R. Bott, S. Gitler and I. M. James, Lectures on Algebraic and Differential Topology. V, 174 pages. 1972. DM 20,-

Vol. 280: Conference on the Theory of Ordinary and Partial Differential Equations. Edited by W. N. Everitt and B. D. Sleeman. XV, 367 pages. 1972. DM 29,-

Vol. 281: Coherence in Categories. Edited by S. Mac Lane. VII, 235 pages. 1972. DM 22,-

Vol. 282: W. Klingenberg und P. Flaschel, Riemannsche Hilbertmannigfaltigkeiten. Periodische Geodätische. VII, 211 Seiten. 1972. DM 22,-

Vol. 283: L. Illusie, Complexe Cotangent et Déformations II. VII, 304 pages. 1972. DM 27,-

Vol. 284: P. A. Meyer, Martingales and Stochastic Integrals I. VI, 89 pages. 1972. DM 18,-

Vol. 285: P. de la Harpe, Classical Banach-Lie Algebras and Banach-Lie Groups of Operators in Hilbert Space. III, 160 pages. 1972. DM 18,-

Vol. 286: S. Murakami, On Automorphisms of Siegel Domains. V, 95 pages. 1972. DM 18,-

Vol. 287: Hyperfunctions and Pseudo-Differential Equations. Edited by H. Komatsu. VII, 529 pages. 1973. DM 40,-

Vol. 288: Groupes de Monodromie en Géométrie Algébrique. (SGA 7 I). Dirigé par A. Grothendieck. IX, 523 pages. 1972. DM 55,-

Vol. 289: B. Fuglede, Finely Harmonic Functions. III, 188. 1972. DM 20,-

Vol. 290: D. B. Zagier, Equivariant Pontrjagin Classes and Applications to Orbit Spaces. IX, 130 pages. 1972. DM 18,-

Vol. 291: P. Orlik, Seifert Manifolds. VIII, 155 pages. 1972. DM 18,-

Vol. 292: W. D. Wallis, A. P. Street and J. S. Wallis, Combinatorics: Room Squares, Sum-Free Sets, Hadamard Matrices. V, 508 pages. 1972. DM 55,-

Vol. 293: R. A. DeVore, The Approximation of Continuous Functions by Positive Linear Operators. VIII, 289 pages. 1972. DM 27,-

Vol. 294: Stability of Stochastic Dynamical Systems. Edited by R. F. Curtain. IX, 332 pages. 1972. DM 29,-

Vol. 295: C. Dellacherie, Ensembles Analytiques Capacités. Mesures de Hausdorff. XII, 123 pages. 1972. DM 18,-

Vol. 296: Probability and Information Theory II. Edited by M. Behara, K. Krickeberg and J. Wolfowitz. V, 223 pages. 1973. DM 22,-

Vol. 297: J. Garnett, Analytic Capacity and Measure. IV, 138 pages. 1972. DM 18,-

Vol. 298: Proceedings of the Second Conference on Compact Transformation Groups. Part 1. XIII, 453 pages. 1972. DM 35,-

Vol. 299: Proceedings of the Second Conference on Compact Transformation Groups. Part 2. XIV, 327 pages. 1972. DM 29,-

Vol. 300: P. Eymard, Moyennes Invariantes et Représentations Unitaires. II, 113 pages. 1972. DM 18,-

Vol. 301: F. Pittnauer, Vorlesungen über asymptotische Reihen. VI, 186 Seiten. 1972. DM 18,-

Vol. 302: M. Demazure, Lectures on p-Divisible Groups. V, 98 pages. 1972. DM 18,-

Vol. 303: Graph Theory and Applications. Edited by Y. Alavi, D. R. Lick and A. T. White. IX, 329 pages. 1972. DM 26,-

Vol. 304: A. K. Bousfield and D. M. Kan, Homotopy Limits, Completions and Localizations. V, 348 pages. 1972. DM 29,-

Vol. 305: Théorie des Topos et Cohomologie Etale des Schémas. Tome 3. (SGA 4). Dirigé par M. Artin, A. Grothendieck et J. L. Verdier. VI, 640 pages. 1973. DM 55,-

Vol. 306: H. Luckhardt, Extensional Gödel Functional Interpretation. VI, 161 pages. 1973. DM 20,-

Vol. 307: J. L. Bretagnolle, S. D. Chatterji et P.-A. Meyer, Ecole d'été de Probabilités: Processus Stochastiques. VI, 198 pages. 1973. DM 22,-

Vol. 308: D. Knutson, λ-Rings and the Representation Theory of the Symmetric Group. IV, 203 pages. 1973. DM 22,-

Vol. 309: D. H. Sattinger, Topics in Stability and Bifurcation Theory. VI, 190 pages. 1973. DM 20,-

Vol. 310: B. Iversen, Generic Local Structure of the Morphisms in Commutative Algebra. IV, 108 pages. 1973. DM 18,-

Vol. 311: Conference on Commutative Algebra. Edited by J. W. Brewer and E. A. Rutter. VII, 251 pages. 1973. DM 24,-

Vol. 312: Symposium on Ordinary Differential Equations. Edited by W. A. Harris, Jr. and Y. Sibuya. VIII, 204 pages. 1973. DM 22,-

Vol. 313: K. Jörgens and J. Weidmann, Spectral Properties of Hamiltonian Operators. III, 140 pages. 1973. DM 18,-

Vol. 314: M. Deuring, Lectures on the Theory of Algebraic Functions of One Variable. VI, 151 pages. 1973. DM 18,-

Vol. 315: K. Bichteler, Integration Theory (with Special Attention to Vector Measures). VI, 357 pages. 1973. DM 29,-

Vol. 316: Symposium on Non-Well-Posed Problems and Logarithmic Convexity. Edited by R. J. Knops. V, 176 pages. 1973. DM 20,-

Vol. 317: Séminaire Bourbaki – vol. 1971/72. Exposés 400–417. IV, 361 pages. 1973. DM 29,-

Vol. 318: Recent Advances in Topological Dynamics. Edited by A. Beck. VIII, 285 pages. 1973. DM 27,-

Vol. 319: Conference on Group Theory. Edited by R. W. Gatterdam and K. W. Weston. V, 188 pages. 1973. DM 20,-

Vol. 320: Modular Functions of One Variable I. Edited by W. Kuyk. V, 195 pages. 1973. DM 20,-

Vol. 321: Séminaire de Probabilités VII. Edité par P. A. Meyer. VI, 322 pages. 1973. DM 29,-

Vol. 322: Nonlinear Problems in the Physical Sciences and Biology. Edited by I. Stakgold, D. D. Joseph and D. H. Sattinger. VIII, 357 pages. 1973. DM 29,-

Vol. 323: J. L. Lions, Perturbations Singulières dans les Problèmes aux Limites et en Contrôle Optimal. XII, 645 pages. 1973. DM 46,-

Vol. 324: K. Kreith, Oscillation Theory. VI, 109 pages. 1973. DM 18,-

Vol. 325: C.-C. Chou, La Transformation de Fourier Complexe et L'Equation de Convolution. IX, 137 pages. 1973. DM 18,-

Vol. 326: A. Robert, Elliptic Curves. VIII, 264 pages. 1973. DM 24,-

Vol. 327: E. Matlis, One-Dimensional Cohen-Macaulay Rings. XII, 157 pages. 1973. DM 20,-

Vol. 328: J. R. Büchi and D. Siefkes, The Monadic Second Order Theory of All Countable Ordinals. VI, 217 pages. 1973. DM 22,-

Vol. 329: W. Trebels, Multipliers for (C, α)-Bounded Fourier Expansions in Banach Spaces and Approximation Theory. VII, 103 pages. 1973. DM 18,-

Vol. 330: Proceedings of the Second Japan-USSR Symposium on Probability Theory. Edited by G. Maruyama and Yu. V. Prokhorov. VI, 550 pages. 1973. DM 40,-

Vol. 331: Summer School on Topological Vector Spaces. Edited by L. Waelbroeck. VI, 226 pages. 1973. DM 22,-

Vol. 332: Séminaire Pierre Lelong (Analyse) Année 1971-1972. V, 131 pages. 1973. DM 18,-

Vol. 333: Numerische, insbesondere approximationstheoretische Behandlung von Funktionalgleichungen. Herausgegeben von R. Ansorge und W. Törnig. VI, 296 Seiten. 1973. DM 27,-

Vol. 334: F. Schweiger, The Metrical Theory of Jacobi-Perron Algorithm. V, 111 pages. 1973. DM 18,-

Vol. 335: H. Huck, R. Roitzsch, U. Simon, W. Vortisch, R. Walden, B. Wegner und W. Wendland, Beweismethoden der Differentialgeometrie im Großen. IX, 159 Seiten. 1973. DM 20,-

Vol. 336: L'Analyse Harmonique dans le Domaine Complexe. Edité par E. J. Akutowicz. VIII, 169 pages. 1973. DM 20,-

Vol. 337: Cambridge Summer School in Mathematical Logic. Edited by A. R. D. Mathias and H. Rogers. IX, 660 pages. 1973. DM 46,-

Vol. 338: J. Lindenstrauss and L. Tzafriri, Classical Banach Spaces. IX, 243 pages. 1973. DM 24,-

Vol. 339: G. Kempf, F. Knudsen, D. Mumford and B. Saint-Donat, Toroidal Embeddings I. VIII, 209 pages. 1973. DM 22,-

Vol. 340: Groupes de Monodromie en Géométrie Algébrique. (SGA 7 II). Par P. Deligne et N. Katz. X, 438 pages. 1973. DM 44,-

Vol. 341: Algebraic K-Theory I, Higher K-Theories. Edited by H. Bass. XV, 335 pages. 1973. DM 29,-

Vol. 342: Algebraic K-Theory II, "Classical" Algebraic K-Theory, and Connections with Arithmetic. Edited by H. Bass. XV, 527 pages. 1973. DM 40,-

Vol. 343: Algebraic K-Theory III, Hermitian K-Theory and Geometric Applications. Edited by H. Bass. XV, 572 pages. 1973. DM 40,-

Vol. 344: A. S. Troelstra (Editor), Metamathematical Investigation of Intuitionistic Arithmetic and Analysis. XVII, 485 pages. 1973. DM 38,-

Vol. 345: Proceedings of a Conference on Operator Theory. Edited by P. A. Fillmore. VI, 228 pages. 1973. DM 22,-

Vol. 346: Fučík et al., Spectral Analysis of Nonlinear Operators. II, 287 pages. 1973. DM 26,-

Vol. 347: J. M. Boardman and R. M. Vogt, Homotopy Invariant Algebraic Structures on Topological Spaces. X, 257 pages. 1973. DM 24,-

Vol. 348: A. M. Mathai and R. K. Saxena, Generalized Hypergeometric Functions with Applications in Statistics and Physical Sciences. VII, 314 pages. 1973. DM 26,-

Vol. 349: Modular Functions of One Variable II. Edited by W. Kuyk and P. Deligne. V, 598 pages. 1973. DM 38,-

Vol. 350: Modular Functions of One Variable III. Edited by W. Kuyk and J.-P. Serre. V, 350 pages. 1973. DM 26,-

Vol. 351: H. Tachikawa, Quasi-Frobenius Rings and Generalizations. XI, 172 pages. 1973. DM 20,-

Vol. 352: J. D. Fay, Theta Functions on Riemann Surfaces. V, 137 pages. 1973. DM 18,-

Vol. 353: Proceedings of the Conference on Orders, Group Rings and Related Topics. Organized by J. S. Hsia, M. L. Madan and T. G. Ralley. X, 224 pages. 1973. DM 22,-

Vol. 354: K. J. Devlin, Aspects of Constructibility. XII, 240 pages. 1973. DM 24,-

Vol. 355: M. Sion, A Theory of Semigroup Valued Measures. V, 140 pages. 1973. DM 18,-

Vol. 356: W. L. J. van der Kallen, Infinitesimally Central-Extensions of Chevalley Groups. VII, 147 pages. 1973. DM 18,-

Vol. 357: W. Borho, P. Gabriel und R. Rentschler, Primideale in Einhüllenden auflösbarer Lie-Algebren. V, 182 Seiten. 1973. DM 20,-

Vol. 358: F. L. Williams, Tensor Products of Principal Series Representations. VI, 132 pages. 1973. DM 18,-

Vol. 359: U. Stammbach, Homology in Group Theory. VIII, 183 pages. 1973. DM 20,-

Vol. 360: W. J. Padgett and R. L. Taylor, Laws of Large Numbers for Normed Linear Spaces and Certain Fréchet Spaces. VI, 111 pages. 1973. DM 18,-

Vol. 361: J. W. Schutz, Foundations of Special Relativity: Kinematic Axioms for Minkowski Space Time. XX, 314 pages. 1973. DM 26,-

Vol. 362: Proceedings of the Conference on Numerical Solution of Ordinary Differential Equations. Edited by D. Bettis. VIII, 490 pages. 1974. DM 34,-

Vol. 363: Conference on the Numerical Solution of Differential Equations. Edited by G. A. Watson. IX, 221 pages. 1974. DM 20,-

Vol. 364: Proceedings on Infinite Dimensional Holomorphy. Edited by T. L. Hayden and T. J. Suffridge. VII, 212 pages. 1974. DM 20,-

Vol. 365: R. P. Gilbert, Constructive Methods for Elliptic Equations. VII, 397 pages. 1974. DM 26,-

Vol. 366: R. Steinberg, Conjugacy Classes in Algebraic Groups (Notes by V. V. Deodhar). VI, 159 pages. 1974. DM 18,-

Vol. 367: K. Langmann und W. Lütkebohmert, Cousinverteilungen und Fortsetzungssätze. VI, 151 Seiten. 1974. DM 16,-

Vol. 368: R. J. Milgram, Unstable Homotopy from the Stable Point of View. V, 109 pages. 1974. DM 16,-

Vol. 369: Victoria Symposium on Nonstandard Analysis. Edited by A. Hurd and P. Loeb. XVIII, 339 pages. 1974. DM 26,-

Vol. 370: B. Mazur and W. Messing, Universal Extensions an One Dimensional Crystalline Cohomology. VII, 134 pages. 197 DM 16,-

Vol. 371: V. Poenaru, Analyse Différentielle. V, 228 pages. 197 DM 20,-

Vol. 372: Proceedings of the Second International Conferenc on the Theory of Groups 1973. Edited by M. F. Newman. VI 740 pages. 1974. DM 48,-

Vol. 373: A. E. R. Woodcock and T. Poston, A Geometrical Stud of the Elementary Catastrophes. V, 257 pages. 1974. DM 22,

Vol. 374: S. Yamamuro, Differential Calculus in Topologic Linear Spaces. IV, 179 pages. 1974. DM 18,-

Vol. 375: Topology Conference 1973. Edited by R. F. Dickma Jr. and P. Fletcher. X, 283 pages. 1974. DM 24,-

Vol. 376: D. B. Osteyee and I. J. Good, Information, Weight o Evidence, the Singularity between Probability Measures an Signal Detection. XI, 156 pages. 1974. DM 16.-

Vol. 377: A. M. Fink, Almost Periodic Differential Equation VIII, 336 pages. 1974. DM 26,-

Vol. 378: TOPO 72 - General Topology and its Application Proceedings 1972. Edited by R. Alò, R. W. Heath and J. Nagat XIV, 651 pages. 1974. DM 50,-

Vol. 379: A. Badrikian et S. Chevet, Mesures Cylindrique Espaces de Wiener et Fonctions Aléatoires Gaussiennes. X 383 pages. 1974. DM 32,-

Vol. 380: M. Petrich, Rings and Semigroups. VIII, 182 page 1974. DM 18,-

Vol. 381: Séminaire de Probabilités VIII. Edité par P. A. Meye IX, 354 pages. 1974. DM 32,-

Vol. 382: J. H. van Lint, Combinatorial Theory Seminar Ein hoven University of Technology. VI, 131 pages. 1974. DM 18,

Vol. 383: Séminaire Bourbaki - vol. 1972/73. Exposés 418-43 IV, 334 pages. 1974. DM 30,-

Vol. 384: Functional Analysis and Applications, Proceeding 1972. Edited by L. Nachbin. V, 270 pages. 1974. DM 22,-

Vol. 385: J. Douglas Jr. and T. Dupont, Collocation Methods f Parabolic Equations in a Single Space Variable (Based on C Piecewise-Polynomial Spaces). V, 147 pages. 1974. DM 16

Vol. 386: J. Tits, Buildings of Spherical Type and Finite B Pairs. IX, 299 pages. 1974. DM 24,-

Vol. 387: C. P. Bruter, Eléments de la Théorie des Matroïde V, 138 pages. 1974. DM 18,-

Vol. 388: R. L. Lipsman, Group Representations. X, 166 page 1974. DM 20,-

Vol. 389: M.-A. Knus et M. Ojanguren, Théorie de la Descen et Algèbres d' Azumaya. IV, 163 pages. 1974. DM 20,-

Vol. 390: P. A. Meyer, P. Priouret et F. Spitzer, Ecole d'Eté Probabilités de Saint-Flour III - 1973. Edité par A. Badriki et P.-L. Hennequin. VIII, 189 pages. 1974. DM 20,-

Vol. 391: J. Gray, Formal Category Theory: Adjointness for Categories. XII, 282 pages. 1974. DM 24,-

Vol. 392: Géométrie Différentielle, Colloque, Santiago Compostela, Espagne 1972. Edité par E. Vidal. VI, 225 page 1974. DM 20,-

Vol. 393: G. Wassermann, Stability of Unfoldings. IX, 164 page 1974. DM 20,-

Vol. 394: W. M. Patterson 3rd, Iterative Methods for the Soluti of a Linear Operator Equation in Hilbert Space - A Surv III, 183 pages. 1974. DM 20,-

Vol. 395: Numerische Behandlung nichtlinearer Integrodiffere tial- und Differentialgleichungen. Tagung 1973. Herausgegeb von R. Ansorge und W. Törnig. VII, 313 Seiten. 1974. DM 2

Vol. 396: K. H. Hofmann, M. Mislove and A. Stralka, The Pont agin Duality of Compact O-Dimensional Semilattices and Applications. XVI, 122 pages. 1974. DM 18,-

Vol. 397: T. Yamada, The Schur Subgroup of the Brauer Gro V, 159 pages. 1974. DM 18,-

Vol. 398: Théories de l'Information, Actes des Rencontres Marseille-Luminy, 1973. Edité par J. Kampé de Fériet et C. Pica XII, 201 pages. 1974. DM 23,-